T0205862

Human–Automation Interaction Design

Transportation Human Factors: Aerospace, Aviation, Maritime, Rail, and Road Series

Series Editor: Professor Neville A. Stanton,
University of Southampton, UK

Automobile Automation
Distributed Cognition on the Road
Victoria A. Banks and Neville A. Stanton

Eco-Driving
From Strategies to Interfaces
Rich C. McIlroy and Neville A. Stanton

Driver Reactions to Automated Vehicles
A Practical Guide for Design and Evaluation
Alexander Eriksson and Neville A. Stanton

Systems Thinking in Practice
Applications of the Event Analysis of Systemic Teamwork Method
Paul Salmon, Neville A. Stanton, and Guy Walker

Individual Latent Error Detection (I-LED)
Making Systems Safer
Justin R.E. Saward and Neville A. Stanton

Driver Distraction
A Sociotechnical Systems Approach
Katie J. Parnell, Neville A. Stanton, and Katherine L. Plant

Designing Interaction and Interfaces for Automated Vehicles
User-Centred Ecological Design and Testing
Neville A. Stanton, Kirsten M.A. Revell, and Patrick Langdon

Human–Automation Interaction Design
Developing a Vehicle Automation Assistant
Jediah R. Clark, Neville A. Stanton, and Kirsten M.A. Revell

For more information about this series, please visit: www.crcpress.com/
Transportation-Human-Factors/book-series/CRCTRNHUMFACAER

Human–Automation Interaction Design

Developing a Vehicle Automation Assistant

Jediah R. Clark, Neville A. Stanton
and Kirsten M.A. Revell

CRC Press
Taylor & Francis Group
Boca Raton London

CRC Press is an imprint of the
Taylor & Francis Group, an **informa** business

First edition published 2022
by CRC Press
6000 Broken Sound Parkway NW, Suite 300, Boca Raton, FL 33487–2742

and by CRC Press
2 Park Square, Milton Park, Abingdon, Oxon, OX14 4RN

CRC Press is an imprint of Taylor & Francis Group, LLC

Reasonable efforts have been made to publish reliable data and information, but the author and publisher cannot assume responsibility for the validity of all materials or the consequences of their use. The authors and publishers have attempted to trace the copyright holders of all material reproduced in this publication and apologize to copyright holders if permission to publish in this form has not been obtained. If any copyright material has not been acknowledged please write and let us know so we may rectify in any future reprint.

Trademark notice: Product or corporate names may be trademarks or registered trademarks and are used only for identification and explanation without intent to infringe.

ISBN: 978-1-032-10161-3 (hbk)
ISBN: 978-1-032-10162-0 (pbk)
ISBN: 978-1-003-21396-3 (ebk)

DOI: 10.1201/9781003213963

Typeset in Times
by Apex CoVantage, LLC

Contents

Section I Scoping the Issues and Solutions that Other Domains Face with Task Continuity

Section II   Pilot Testing These Concepts in Automated Driving

Section IV   *Testing and Validating a Novel Prototype*

Preface

Automated technology is fast paced. Increased computing power and global pressure to address shortcomings and efficiency in transportation has led to an arms race to spearhead innovation and product development. For many of us, this eventuality will likely be a positive change; however, such a fast pace means that it is all too easy to forget about the end user and how actions and behaviors are influenced by these technological developments. This book addresses this issue by keeping the user and society as the primary focus and by developing new ways for users to interact with technology that promotes safe operation and appropriate use.

As will be discussed in this book, the issues within shared-control automated technology are complex. Varying contexts, environmental pressures, and user requirements can make the concept of situation awareness in automated vehicle operation a tricky one to tackle. This book was conceived to address this issue by developing a flexible, usable, and safety-oriented user interaction that provides the driver with crucial information in a variety of scenarios. Much of this work is inspired by how humans naturally communicate to one another, a concept that has decreased in popularity over the past decade. However, as automated and autonomous technology develops, this approach may in fact become more relevant to advancing automated vehicle technology. This book promotes and modernizes the idea of cooperative automation to become more pragmatic and adhere to the modern models of situation awareness and demonstrates that wherever there are acting agents communicating, there is a lesson to be learned.

This book serves not only to provide the reader with an insight into what the main issues are in future automated vehicles but to provide solutions, demonstrations, and an example road map of how to take theoretical human–machine interface concepts and apply them to practice. The link between an idea and a working prototype can be hard to create; therefore, the methods introduced and demonstrated in this book may help you in bridging your own gaps in whichever domain you find yourself working. The book consists of four sections and outlines the journey from 1) scoping the issues and solutions that other domains face with task continuity, 2) pilot testing these concepts in automated driving, 3) designing new interfaces and interactions for automated vehicle communication, and 4) testing and validating a novel prototype.

Dr Jediah R. Clark, PhD
University of Southampton

Acknowledgements

The work within this book was conducted as part of the UK's nationwide project 'Towards Autonomy: Smart and Connected Control', forming a partnership between the EPSRC, Jaguar Land Rover, and eight universities consisting of the University of Southampton, University of Cambridge, University of Edinburgh, University of Birmingham, Heriot-Watt University, Cranfield University, University of Surrey, and the University of Warwick. The subproject for this work, 'Human Interaction: Designing Autonomy in Vehicles' (HI:DAVe), addressed the human factors issues within the project with the goal of improving interactions and interfaces for safety-critical operation of automated vehicles. HI:DAVe consisted of the University of Southampton in partnership with the University of Cambridge from which we acknowledge the contributions of Dr Patrick Langdon, Dr Mike Bradley, and Nermin Caber toward making this work possible. Additionally, we acknowledge the following colleagues from the Human Factors Engineering team at the University of Southampton for their academic support: James Brown, Joy Richardson, Dr Jisun Kim, and Dr Sylwia Kaduk. The authors also acknowledge the support from the technical partners at Jaguar Land Rover who provided industrial insights into the interface solutions provided in this book: Dr Lee Skrypchuk, Simon Thompson, and Jim O'Donoghue.

Research is made more effective with institutional support networks and the participation of the public in user-participation trials and interviews. Thank you to those who contributed in this way and worked with us to ensure that this research was a success.

This work is devoted to our global societies and end users of automated technology. Their safety and well-being should never be overseen in favor of rapid technological development. The future of safe autonomy relies on smart design and user participation, and we at the Human Factors Engineering team at the University of Southampton are delighted to present our contributions toward this ideal.

Authors

Dr Jediah R. Clark, PhD, is a Research Fellow in Human–Automation Interaction. He has a PhD degree in Human Factors and an MSc degree in Research Methods from the University of Southampton and a BSc degree in Psychology from Cardiff Metropolitan University. His research interests include interface and interaction design, and communication protocol design for the purposes of raising system situation awareness and improving the calibration of trust. Dr Clark was awarded the Southampton University Doctoral College Enterprise Award in 2019 and 2020 for his contributions to the design of automated vehicle interfaces in industry. He has worked on automated vehicle human–machine interface (HMI) design for 5 years and has also worked in the domains of defense and autonomous drone operation. Dr Clark believes that autonomous systems should be tailored and carefully designed to the user population to ensure that these complex systems work with humanity rather than against it.

Professor Neville A. Stanton, PhD, DSc, is a Chartered Psychologist, Chartered Ergonomist, and Chartered Engineer. He has recently retired from the Chair in Human Factors Engineering in the Faculty of Engineering and the Environment at the University of Southampton in the UK. He has degrees in Occupational Psychology, Applied Psychology, and Human Factors Engineering and has worked at the Universities of Aston, Brunel, Cornell, and MIT. His research interests include modeling, predicting, analyzing, and evaluating human performance in systems as well as designing the interfaces and interaction between humans and technology. Professor Stanton has worked on design of automobiles, aircraft, ships, and control rooms over the past 30 years on a variety of automation projects. He has published 50 books and more than 400 peer-reviewed journal papers on Ergonomics and Human Factors. In 1998, he was presented with the Institution of Electrical Engineers Divisional Premium Award for research into System Safety. The Institute of Ergonomics and Human Factors in the UK awarded him The Otto Edholm Medal in 2001, The President's Medal in 2008 and 2018, The Sir Frederic Bartlett Medal in 2012, and The William Floyd Medal in 2019 for his contributions to basic and applied ergonomics research. The Royal Aeronautical Society awarded him and his colleagues the Hodgson Prize in 2006 for research on design-induced flight-deck error published in *The Aeronautical Journal*. The University of Southampton has awarded him a Doctor of Science in 2014 for his sustained contribution to the development and validation of Human Factors methods.

Dr Kirsten M. A. Revell has, for a number of years, led the autonomous vehicle domain within the Human Factors Engineering team as a Research Fellow in the Transportation Research Group (TRG) in the Faculty of Engineering and the Environment at the University of Southampton. She has degrees in both Psychology and Industrial Design from Exeter and Brunel University London, respectively, as well as a PhD in Human Factors for the design of behavior change from the

University of Southampton. She has previously worked in Microsoft Ltd. and has also conducted research in military, domestic energy, rail, and aviation domains, collaborating extensively with government bodies and industry. She is a member of the Human Factors Sustainable Development Technical Committee and has been jointly awarded the Annual Aviation Safety Prize by the Honourable Company of Air Pilots and the Air Pilots Trust. Dr Revell passionately believes that Human Factors can offer solutions to the critical global issues we face today. Her current focus is on using Human Factors to show how systems shape lives, highlighting where change is needed to promote an inclusive and sustainable world, with a particular interest in gender equity.

1 Introduction

1.1 BACKGROUND

A driverless future where automated systems are able to control a road vehicle and make strategic decisions for the driver promises a wide range of benefits including a reduction in road traffic accidents, an increase in users' free time, and an increase in traffic and fuel efficiency (DFT, 2015). This automated future is one that has been embraced by the UK government (DFT, 2015) and has marked the start of a race among manufacturers to roll out models with increasingly sophisticated automated features (Fagnant & Kockelman, 2015). Given the myriad of benefits and the optimism showed by many, a critical and skeptical approach has been advised by the UK government and independent institutes to ensure that safety standards are adhered to (GOV, 2017).

The levels of automation put forward by SAE international (SAE, 2016) attempt to tighten the discourse surrounding automated vehicles (AVs) and their operational capacity by categorizing automated features into six discrete levels outlined in Table 1.1.

These levels outline vehicles with no automated features (level 0) up to vehicles that require no human driver inputs to operate effectively (level 5). Between these two extremes, four levels of automation represent varying degrees of human driver tasks and responsibilities—each requiring both driver and automation to perform certain tasks at certain intervals. These levels, due to issues surrounding shared control and responsibility, exhibit novel vulnerabilities. For example, incident reports cite driver distraction, overreliance, and human error as being the central cause of collisions in AVs that require monitoring (level 2 automation; Banks et al., 2018; BBC, 2020; Stanton et al., 2019; SAE, 2016). Despite this, increasingly 'sophisticated' AVs are being made available to the public with the driver being further removed from aspects of the driving task. Beta-test AVs are currently equipped with level 2 automation—lateral and longitudinal automation that requires the human driver to monitor the driving task in the case of an emergency. However, level 3 AVs such as Audi's A8 model equipped with traffic jam assist are available to citizens of states where level 3 AVs are road-legal. Level 3 AVs require transitions in control and responsibility as a result of breaching an operational limit (such as loss of central reservation detection in the Audi A8; Audi, 2018; SAE, 2016). These vehicles differ from their level 2 predecessors in that they allow the driver to take part in non-driving-related tasks while automation is in control (SAE, 2016). Level 4, in extension, may feature control transfers although level 4 AVs are assumed to not require falling back to the driver in the case of a breach of operational safety.

The recurring feature of requiring control transitions in shared-control AVs has been identified as contributing toward novel vulnerabilities such as a reduction in situation awareness (Endsley, 1995; Sarter & Woods, 1992, 1995), deterioration

DOI: 10.1201/9781003213963-1

TABLE 1.1
SAE Levels of Automation (SAE, 2016)

Level	Name	Control Executor	Monitoring Environment	Fallback Performance	Capability (modes)
0	No automation	Driver	Driver	Driver	n/a
1	Driver assistance	Driver and system	Driver	Driver	Some
2	Partial automation	System	Driver	Driver	Some
3	Conditional automation	System	System	Driver	Some
4	High automation	System	System	System	Some
5	Full automation	System	System	System	All

Source: Summary adapted from SAE (2016)

in attentional resources (Young & Stanton, 2002a, 2002b), mode error (Norman, 2015; Sarter & Woods, 1995), deskilling (Bainbridge, 1983), and lack of calibration in trust (Koo et al., 2015; Lee & See, 2004). Other factors arising from an increase in driver–automation interaction include the acceptance and usability of the technology (NSAI, 2018; Nwiabu & Adeyanju, 2012; Ponsa et al., 2009; Schieben et al., 2011).

AV technology is progressing quickly, and research must keep up with public and manufacturing demands. As legality of level 3 AVs is granted across the world, issues introduced by shared control and responsibility must be addressed to reduce fatalities and collisions while ensuring that the technology benefits the user with regards to usability and acceptance. This book considers such implications and develops foundations and design solutions for human–machine interfaces (HMIs) to optimize these human factors outcomes in level 3 and 4 AVs as a collective.

1.2   RESEARCH MOTIVATION

Ensuring that developments in automated technology are implemented safely allows manufacturers and the public to benefit from positive outcomes while ensuring that novel vulnerabilities are minimized. By conducting research into its safe development, a driverless future is more likely to be beneficial to societies and becomes a technology that should not be feared but embraced. The content within this book was acquired as part of a broader research project, Human Interaction: Designing Autonomy in Vehicles (HI:DAVe), which in turn was part of a nationwide research program, Towards Autonomy: Smart and Connected Control, funded by both Jaguar Land Rover and the EPSRC. HI:DAVe addressed the human–machine interfaces (HMIs) in level 3 and 4 AVs, giving particular attention to the situations where a takeover request is made as a result of a critical or noncritical operational/design violation (e.g., upcoming geographical boundary; SAE, 2016).

HMIs are central to the solutions posited to reduce vulnerabilities in shared-control AVs as they allow for the driving system to relay information between both driver and vehicle. As driver and automation have distinct roles in the future of AVs, the importance of effective HMIs is ever increasing. To inform design, this book draws upon concepts that aim to improve interaction from theory (distributed situation awareness—DSA; Stanton et al., 2006 & joint activity (JA) framework; Bradshaw et al., 2009; Clark, 1996; Klein et al., 2004, 2005) and practice (shift handover in human teams; e.g., Kerr, 2002). This is with a view of generating a novel HMI design that relays important information to the driver to improve safety following transitions of control while maximizing usability and optimizing trust and workload. Further, this book illustrates an example design lifecycle for human factors practitioners to draw inspiration from, progressing as follows: scoping, piloting, designing, and finally prototype testing, with each chapter being a part of a step in this progression.

The book approaches the issue of 'handover' (defined in this book as the transition of control from vehicle to driver) in an innovative way. The majority of previous research in level 3 AV technology is primarily concerned with transitions of control in response to emergency situations. Transitions of control should be central in discussions on how to improve human–automation interaction in level 3 AVs; however, this book acknowledges that this is an oversimplification of the vulnerabilities in level 3 AVs as handovers may be initiated by either party in response to a variety of events (Mirnig et al., 2017). Further, knowledge on how transitions should occur during nonemergency scenarios is limited, and design solutions that attempt to optimize outcomes are yet to be provided. This book, therefore, provides a unique perspective on the AV handover task by integrating communicative concepts found in other domains to improve communication throughout the automated cycle while ensuring that communication can be made more efficient and tailored to the driver by applying concepts of distributed situation awareness. In doing so, this book contributes to the body of knowledge on how to design HMIs to improve human factors outcomes in AVs requiring control transitions. Improving communication for level 3 vehicles may be a priority; however, findings can translate to automated levels that require both driver and automation to fulfill specific roles in the driving task. Findings from this book can be readily applied to level 4 vehicles as both level 3 and 4 AVs have the potential for human and automation to transfer control and responsibility between one another. It is therefore hoped that the output provided here will inform future HMI design in AVs regardless of specific target context.

1.3   RESEARCH OUTCOMES AND HYPOTHESES

This book aims to produce a handover interface design that improves and optimizes a wide range of experimental outcomes related to human performance: safety, situation awareness, trust, workload, acceptance, and usability. The work addresses many outcomes, as reducing the issue to a single outcome may not provide suitable solutions due to potential trade-offs and optimization problems. The design will achieve this by aligning with preexisting concepts in human team shift-handover practices, communicative principles from human–machine teamwork and the theory of distributed situation awareness. The HMI design will be generated through a four-step method of scoping the field, piloting preliminary concepts, designing a solution, and testing prototype HMIs.

1.3.1 RESEARCH OUTCOMES

Primary Research Outcome

The introduction of more sophisticated automated systems leads to a greater requirement for information to be optimally exchanged between vehicle and driver. The overarching aim of this book is to provide novel solutions inspired by communication literature and rigorous prototype development. Therefore, the primary outcome of this book is:

- To provide an HMI design solution that improves coordination between driver and automation in level 3 and 4 AVs during all phases of a journey.

Secondary Research Outcomes

As an extension of the primary research outcome, insights will be provided for theoretical and methodological approaches to AV design. In particular, the secondary outcomes of this book will be to:

- Provide insight into how communicative concepts (Clark, 1996; Klein et al., 2004, 2005; Bradshaw et al., 2009) and distributed situation awareness (Stanton et al., 2006, 2017b) can be applied to level 3 and 4 AV HMI design.
- Demonstrate how a four-step approach to human factors design can be used to address multiple domain values.
- Provide findings that show how driver demographics may affect driver requirements for level 3 and 4 AV interaction.

1.3.2 RESEARCH HYPOTHESES

Overall Hypothesis

To achieve the primary outcome, the overall hypothesis addresses previous work on how humans communicate with other humans and machines in a variety of high-risk domains. These theoretical frameworks provide a foundation for improving communication in level 3 and 4 AVs. Therefore, the overall hypothesis for this book is as follows:

- Provide findings that show how driver demographics may affect driver requirements for level 3 and 4 AV interaction.

Sub-Hypotheses

Multiple human–automation interaction outcomes were identified throughout this book to measure the impact of introducing novel concepts to level 3 and 4 interaction. During each testing stage of the design pathway, the novel concepts introduced throughout this book are tested via these outcomes to measure how well design recommendations address the issues level 3 and 4 AV interaction face:

- As transitions may lead to breakdowns in physical control and a reduction in situation awareness (Eriksson & Stanton, 2017b; Merat & Jamson, 2009; Stanton et al., 2006; Stanton et al., 2017b)—'the novel design will improve safe operation of the vehicle (e.g., lateral and longitudinal stability)'.

- Due to issues surrounding disuse and misuse of automation (Lee & See, 2004), 'the novel design will better optimize trust'.
- Usability interacts with safe operation and acceptance of interactions (Barón & Green, 2018; NSAI, 2018; Nwiabu & Adeyanju, 2012; Ponsa et al., 2009; Schieben et al., 2011). Therefore, a sub-hypothesis will be: 'The novel design will improve usability'.
- Acceptance of automated technology is an issue that requires addressing from both a public and an individual perspective (van der Laan et al., 1997). Therefore, a sub-hypothesis is that: 'The new design will improve acceptance'.

Optimizing workload is an important issue for interaction researchers, as workload can influence task performance. It is difficult to predict whether increasing or decreasing workload is more optimal for driving performance (as too high or too low workload can lead to reductions in cognitive performance) (Young & Stanton, 2002b): the novel design will influence the workload of drivers during AV operation.

1.4 BOOK STRUCTURE

Figure 1.1 illustrates the approach taken to meet the outcomes outlined in Section 1.2. The book can be broken down into four distinct stages: scoping, piloting, designing, and testing. This approach allows for theory to be addressed and tested during the scoping and piloting stages, so that critical questions are answered prior to the design stage. The design and testing stages aim to collate these findings into testable outcomes for AV HMI implementation and provide evidence as to what elements of these interfaces are important and have positive outcomes on human–automation interaction in these vehicles. Addressing this scope will also allow for current models of human–agent collaboration to be improved upon to facilitate the continued and rapid progression of AV technology.

Data will inform readers of how interface designs can be applied to their domain/ research questions and allow for models of both the handover task and human–automation communication to be improved as a result of this work.

1.4.1 Chapter 1—Overview of Book

This chapter provides a brief overview of the design problem and the issues that level 3 and 4 AVs face. It outlines the background, aims of the book, structure, and the contributions to knowledge. It serves as a preface to the rest of the book before Chapter 2 goes into more detail regarding the issues and proposed HMI solutions currently within this increasingly complex domain.

1.4.2 Chapter 2—Automated Vehicles as a Copilot: Setting the Scene for Effective Human–Automation Collaboration

Chapter 2 is concerned with the levels of automation, the vulnerabilities level 3 and 4 AVs present, and introduces literature into effective communication including

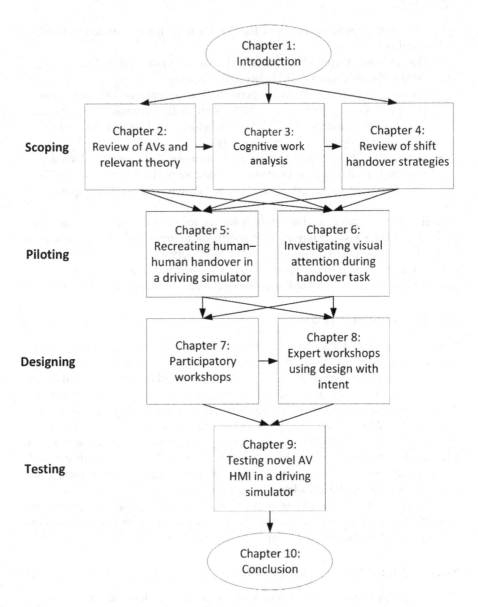

FIGURE 1.1 Book outline—chapters and associated stages.

frameworks such as joint activity and distributed situation awareness. This chapter provides a summary of what aspects of human factors in AV HMIs have been explored previously and how communicative concepts can be used to improve interaction in AVs. This theoretical basis serves as a foundation for analysis of Chapter 3 into the domain's functions and processes and provides a lens in which the remaining chapters can discuss human–automation interaction with regards to improving communication within the system.

1.4.3 CHAPTER 3—COGNITIVE WORK ANALYSIS TO IMPROVE COMMUNICATION IN AV INTERACTIONS

To achieve the design goal, it was deemed important to first scope the domain and understand which affordances and constraints are present within the system of level 3 and 4 automation. This chapter links the outcomes of the book from Chapters 1 and 2 to identify domain values and provide an overview of how these link with physical aspects of the system using cognitive work analysis (CWA; Vicente, 1999)—a versatile human factors method for developing a foundation for making improvements in any working domain. The CWA Design Toolkit (Read et al., 2015) was used to produce a 'Work Domain Analysis', a 'Contextual Activity Template', and a 'Social Organization and Cooperation Analysis'. These three methods together provide insight into the domain processes at work, the planning of tasks that need to be performed, and the allocation of these tasks to both driver and automation. From here, the output of the analysis informs the rest of the book by providing a framework in which designs align to.

1.4.4 CHAPTER 4—REVIEW OF HANDOVER TOOLS AND TECHNIQUES IN HIGH-RISK SHIFT-WORK DOMAINS

Decades of work has been conducted in the continuation of tasks in human teams. Such vast amounts of research may be of use when advising the design of level 3 and 4 AVs. This chapter provides an in-depth review of current handover practices in human teams, with the view of trialing select methods in AV handover in future chapters. Shift handover in domains such as healthcare, aviation, and energy manufacturing has been developed to raise situation awareness and ensure that information transferred is useful to the incoming party. By identifying and assessing the current tools and techniques in these domains, this chapter provides 19 strategies along with examples and a review of how well they meet values of distributed situation awareness. This knowledge contributes to the foundations for handover design in AVs and serves current and new domains with a set of strategies they can implement to ensure safe task continuity.

1.4.5 CHAPTER 5—REPLICATING HUMAN–HUMAN COMMUNICATION IN A VEHICLE: A SIMULATION STUDY

To apply potential handover strategies in Chapter 4 to AV handover interfaces, this chapter recreates the handover task by creating an experiment where two drivers transfer control to one another while communicating vocally in a variety of ways in a driving simulator. Vocal strategies were drawn from Chapter 4 to test how vocal communication could take place in the operation of AVs. Findings on the use of language, information transmitted, method of transmission, workload, usability, acceptance, lateral and longitudinal control following handover are presented to inform which strategies transactions in situation awareness could take and provide recommendations for the use of vocal interfaces in level 3 and 4 AVs.

1.4.6   CHAPTER 6—DIRECTABILITY AND EYE-GAZE: EXPLORING INTERACTIONS BETWEEN VOCAL CUES AND THE USE OF VISUAL DISPLAYS

Chapters 4 and 5 provide findings suggesting that vocal communication can be an effective strategy for communicating information rapidly during handover. This chapter considers which visual HMIs may be best suited to supplement vocal information during the cycle of automation. This chapter analyzes data from a simulated handover task, with an emphasis on where individuals visually attend to during and following the transfer of control. HMIs were split into separate areas of interest and total visual gaze time was analyzed taking into account demographics such as gender, age, time in automation, and premium car ownership. The chapter summarizes the most effective visual HMIs for level 3 and 4 AVs and discusses them in relation to the concept of 'directability' and current models on the market. With this information, future chapters combine these vocal and visual communication strategies to optimize situation awareness transactions and adhere to the principles of joint activity.

1.4.7   CHAPTER 7—PARTICIPATORY WORKSHOPS FOR DESIGNING INTERACTIONS IN AUTOMATED VEHICLES

Ensuring that target users are included in the design lifecycle is regarded as essential for modern human factors design. This chapter represents the first step in developing a testable prototype for level 3 and 4 AV HMIs by presenting findings on what learner, intermediate, and advanced drivers require and suggest for raising situation awareness prior to the handover of control. This chapter explores the process of communicating information, physically transitioning control, and ensuring that users are aware of what needs to be achieved was the primary focus of this study. Drivers discussed solutions and generated schematics illustrating what, and how, information should be communicated at each stage of the handover process. The chapter summarizes these schematics and provides insights into how level 3 and 4 AV HMI design can cater to users' needs while ensuring that system safety is addressed. Further, this study discusses varying requirements for skill levels of driving.

1.4.8   CHAPTER 8—DESIGNING AUTOMATED VEHICLE INTERACTIONS USING DESIGN WITH INTENT

Bringing all previous chapters together, Chapter 8 generates a design concept for level 3 and 4 AV HMIs that focuses on implementing vocal and visual communication while communicating essential information such as system state, capability, and directions. A design workshop was conducted to aid in the converging of ideas that have been built throughout the book. This workshop consisted of five human factors specialists and utilized the design with intent toolkit—101 cards providing concepts that should be addressed for effective user design. As an outcome, this workshop provides design suggestions and an example prototype for implementation in this domain.

1.4.9 CHAPTER 9—VALIDATION AND TESTING OF FINAL INTERACTION DESIGN CONCEPTS FOR AUTOMATED VEHICLES

To validate all work conducted in previous chapters, this chapter generates a handover assistant that is capable of being implemented and evaluated with current technology and is validated in a driving simulator task. The handover assistant is tested against a handover assistant currently available in the AV domain and compares outcomes such as vehicle control, usability, acceptance, trust, communication, and workload. The findings provide a promising outlook on the development of a communicative handover assistant and provides insight into how this could be achieved through vocal and visual HMIs with elements. Notably, the handover assistant tested in this chapter can improve all human factors outcomes without a single outcome becoming degraded. This demonstrates that careful, stepwise HMI design can lead to all-round benefits, given that they adhere to fundamental principles.

1.4.10 CHAPTER 10—CONCLUSIONS

This chapter evaluates the progression made within this research by summarizing the outcomes generated, the success of the method taken, and provides an insight into what current research may be neglecting in the face of new technological developments. The discussion provides a new model for handover and tenets that advise on future handover assistant design to ensure that HMIs in this domain are communicative, collaborative, and take the whole automation cycle into consideration.

1.5 CONTRIBUTION OF KNOWLEDGE

This book provides original practical, theoretical, and methodological contributions for human factors design in level 3 and 4 AVs. The primary outcome of this book is a carefully developed and tested handover assistant that focuses on communicating intentions and state while providing user querying as a way of raising situation awareness. By viewing the handover task as a two-way process where the driver and automation are organized as copilots, the system can distribute tasks effectively while ensuring that both driver and automation communicate intentions, capacity, and safety-critical information. Theoretically, the book discusses current handover assistants in line with the theory of Distributed Situation Awareness and Joint Activity (Klein et al., 2004, 2005; Stanton et al., 2006, 2017b), highlights the requirement for communication to occur throughout the journey, not just the handover, and advises manufacturers to consider how visual and vocal modalities are during each stage of automation. Methodologically, the book demonstrates how HMI design in human factors can be generated on a step-by-step basis, ensuring that theory, domain constraints, user requirements, and real-time human performance are all considered to produce design solutions to challenging human factors issues.

All outcomes combined provide current and future automation researchers and manufacturers with a foundation of how the interaction design process can be approached in conditionally and highly automated vehicles (C/HAVs). Due to the book covering multiple outcomes such as trust, usability, safety, workload, and acceptance, the design pathway presented in this book will be of great use to researchers that require multiple human factors outcomes to be addressed.

1.6 FUTURE DIRECTIONS

This chapter has outlined the aims and objectives of the book and briefly introduces key concepts such the levels of automation, the vulnerabilities introduced by shared-control AVs, literature that may help to address these vulnerabilities, and the pathway proposed to design HMI solutions to address these vulnerabilities. To further analyze the issues within this domain and potential solutions, Chapter 2 provides more depth in the current state of levels of automation, AV vulnerabilities, AV interfaces, and the communicative concepts that could be utilized to address target issues.

Section I

Scoping the Issues and Solutions that Other Domains Face with Task Continuity

2 Vehicle Automation as a Copilot

Setting the Scene for Effective Human– Automation Collaboration

Chapter 2 builds on Chapter 1 by providing an in-depth review of the issues AV development face, a review of the current state of AV HMIs, and a detailed portrayal of the theoretical concepts that may contribute toward improvements in interactions within level 3 and 4 AVs.

Modern driverless vehicles (e.g., Audi A8—Audi, 2019a; Cadillac Super Cruise—Cadillac, 2020; and Tesla Model S—Tesla, 2018) attempt to, among many other proposed benefits, reduce collisions, free-up time and attention from the driving task, as well as optimize traffic flow (DFT, 2015; Maurer et al., 2016; Waldrop, 2015). The future of driverless vehicle technology will require both driver and automation to collaborate with one another to ensure journey success. This chapter considers the issues modern automated vehicles (AVs) face and the challenges that they introduce. It provides an overview of communication and situation awareness theory, and the current state of modern AVs to set the foundations for the target domain and engineering problem.

2.1 LEVELS OF AUTOMATION AND THE HANDOVER

In automobile technology, the levels of automation (SAE, 2016) represent separate avenues that manufacturers can pursue to be part of a driverless future. Each approach comes with its own set of benefits and drawbacks that require consideration during the design process. As defined by SAE (see Table 1.1; SAE, 2016), 'level 5' automation (full automation) involves the removal of driving inputs, so that a human driver is not requested to take control of the vehicle. This stage is thought to require a larger investment from designers and manufacturers to ensure that autonomous vehicles can respond appropriately in any given situation on its route. Inevitably, the collective public attitudes and law will dictate the pace at which these vehicles are developed. The alternative is that vehicle automation is introduced incrementally to allow the public, and technology, to adapt to a driverless future. Conditionally and highly automated vehicles (C/HAVs, levels 3 and 4; Clark et al., 2018; SAE, 2016) represent this next step (levels 3 and 4; SAE, 2016). C/HAVs either expect (level 3) or offer (level 4) the driver to control the vehicle during the journey. This approach may

DOI: 10.1201/9781003213963-3

be more feasible for manufacturers, as there is a greater ability to apply automation selectively to less complex and more predictable scenarios (such as highway driving) while ensuring journey continuity (Kyriakidis et al., 2017). This may be beneficial to the user as a way of improving safety, freeing up more time, and disengaging from a particularly monotonous driving task.

Drivers of level 3 and 4 vehicles are able to take part in 'secondary activities'—tasks that are not directly relevant to driving such as engaging with entertainment or conducting work activities. Counterintuitively, level 3 automation requires the driver to be 'on-hand' in case they need to intervene in response to a design/system violation. With Audi having released its 'traffic jam assist' for the Audi A8 in early 2018, this new era of AV technology is now in motion (Audi, 2019a). Audi's A8 is the first road-legal vehicle that allows the driver to direct their attention away from the driving environment under specific conditions. The Audi A8 is conditional, as it can only activate during a traffic jam under 45 kmph and requires the tracking of a central reservation. If these conditions cannot be met, the driver is notified and expected to withhold the secondary task and takeover control from the automated system—failing to do so could lead to a collision. Therefore, the system is required to conduct a 'handover'—the transition of control from vehicle to driver. When automation conditions are reestablished, the system can then conduct a 'handback'—the transition of control from driver to vehicle. Throughout this book, these terms are used to represent the direction of control transition being discussed.

Level 3 and 4 automation share that automation will be in full control of lateral and longitudinal control of the vehicle; however, level 4 AVs represent a step toward greater autonomy as failing to respond to a takeover request does not put the safety of the system at risk.

2.2 EMERGENT ISSUES IN LEVEL 3 AND 4 AUTOMATION

Although benefits are plentiful for these AVs, due to the distributed nature of tasks for both driver and automation in C/HAVs and the increased likelihood of coordination deficiencies many emergent issues arise such as:

- Reductions in SA—Out-of-the-loop performance where a reduction in situation awareness occurs when an operator is expected to take control after being disconnected from the environment for a set time (Endsley & Kiris, 1995; Heikoop et al., 2016; Stanton & Young, 2000). This may have implications for safety as illustrated across domains that require such handover tasks (e.g., Adamson et al., 1999; Brandenburg & Skottke, 2014; de Carvalho et al., 2012; de Winter et al., 2014; Endsley, 1995; Endsley & Kiris, 1995; Louw et al., 2015; Merat & Jamson, 2009; Patterson et al., 2004; Stanton et al., 2017a).
- Degradations in vehicle control following a driver regaining control (Brandenburg & Skottke, 2014; Eriksson & Stanton, 2017b; Merat & Jamson, 2009).
- Trust may not be calibrated to match capabilities (Lee & See, 2004; Walker et al., 2016), and therefore automation could be misused or disused.

- Mode errors—A human operator may misinterpret mode status and lead to a situation where they are/are not in control of the vehicle at the appropriate time (Stanton et al., 2011). This is illustrated by Bainbridge (1983) and a number of recorded incidents related to mode error (Sarter & Woods, 1992, 1995).
- Increased requirement for interaction—Greater emphasis on other agents performing tasks leads to a greater requirement for interactions and contributes another level of consideration toward a system. When multiple agents are required for task success, coordination is essential to ensure efficient task performance (Salas et al., 2000; Sheridan, 2002).
- Requirement for dynamic and adaptive function allocation (Fuld, 2000; Idris et al., 2016).
- Not adhering to usability principles that focus on user requirements and the nature of interaction which, in turn, reduces effectiveness, efficiency, and satisfaction (Barón & Green, 2018; NSAI, 2018; Nwiabu & Adeyanju, 2012; Ponsa et al., 2009; Schieben et al., 2011).
- Workload may increase as a result of monitoring the automation's performance over time (de Winter et al., 2014; Young & Stanton, 2002b). Low workload (especially when automation is introduced) can have an effect on attentional capacity (Young & Stanton, 2002a, 2002b, 2004).
- Requirement to allocate legal responsibility of the vehicle (Kyriakidis et al., 2017; SAE, 2016).

The handover has been highlighted as the time period where many of these vulnerabilities are likely to manifest themselves (Molesworth & Estival, 2015; Thomas et al., 2013). This is largely due to the main factor affecting safety—the reduction of SA when the driver is expected to take control from an automated system (Brandenburg & Skottke, 2014; Merat & Jamson, 2009; Merat et al., 2012). Handovers are initiated by driver or by vehicle due to a variety of events (McCall et al., 2016). These 'events' can be categorized as being either a *critical event* or *emergency* (e.g., sensor failure, lost track of leading vehicle, and dangerous weather conditions) or a *noncritical event* or *nonemergency event* (e.g., geographical or pre-expected design boundary; Banks & Stanton, 2016; Eriksson & Stanton, 2017a; SAE, 2016; Stanton & Marsden, 1996). Critical events, due to their hazardous nature, require quick intervention from the driver to prevent a potential collision.

Critical handover events do not typically give the driver a time allowance to raise SA prior to the driver taking control. Conversely, when boundaries are predictable (e.g., roadworks and exiting junction on motorway to urban area), time is likely to be more readily available. As Patterson and Woods (2001) state, the handover is a time where the incoming operator must raise SA and have a complete mental model of the situation and anything that has changed. In the case of a noncritical event the handover should take place over a 'comfortable transition time' to ensure adequate time for SA to be raised, as well as being adaptable to the driving context and driver awareness (Eriksson & Stanton, 2017c; Merat et al., 2014; NHTSA, 2013; Walch et al., 2015). For noncritical handovers, there is still much debate over what constitutes as 'comfortable transition time' (Merat et al., 2014) although a study by Eriksson and Stanton (2017a) showed that this ranged between 1.97 and 25.75 s (Mdn = 4.56) when simply asked to takeover with no time restriction. Willemsen et al. (2014) propose that takeover time should be modified based on driver awareness prior to handover. Level 3 AV research typically focuses on emergency scenarios where time criticality is of great importance (Eriksson & Stanton, 2017a); however, planned vehicle-to-driver handover will occur at least once every journey and remains largely unexplored.

2.3   CURRENT STATE OF HANDOVER ASSISTANTS

In their comprehensive review in transition interfaces, Mirnig et al. (2017) outline the current state of transition interfaces regarding a categorization framework. The authors identify contributory work across academia and industry. Notable design specifications from this review include: 1) alerts informing of situation and takeover time (Walch et al., 2015), 2) implementing bimodal (auditory and visual) takeover requests (Walch et al., 2017), 3) exploring multimodal alerts and the effect of direction on takeover performance (Petermeijer et al., 2017a, 2017b), 4) ambient and contextual cues to facilitate takeover (Borojeni et al., 2016), 5) graded takeover request in 'soft takeover request' scenarios (Forster et al., 2016), and 6) multimodal alerts in relation to urgency (Politis et al., 2015).

As an example, Naujoks et al. (2017) provide a prototype handover interface that gives the driver information about the current situation prior to taking control. Elements include the speed of the vehicle, the type of road event that is causing the handover, and distance to the event. The request is displayed in two different ways: 'non-imminent'—banners colored in orange, with a wheel indicating how much time remains for takeover to occur—and 'imminent', banners colored in red and a more urgent message. Their interface recommendation follows concepts related to cooperative perception technology (Naujoks & Neukum, 2014; Naujoks et al., 2014) which involves interfaces feeding real time, event-critical information to the driver to improve safety following a takeover request.

Many other areas related to C/HAV handover have been addressed, such as SA (Merat & Jamson, 2009; Stanton et al., 1997), notifications (Bazilinskyy & de Winter, 2015), time to takeover (Eriksson & Stanton, 2017a; Gold et al., 2017; Young & Stanton, 2007b; Zeeb et al., 2015), effect of demographics (Körber et al., 2016), effect of traffic density (Gold et al., 2016), effect on driver behavior (Merat et al., 2014; Naujoks & Neukum, 2014; Naujoks et al., 2014), distractions (Mok et al., 2015), temporal/complexity constraints (Eriksson et al., 2015), and handover assistants (Eriksson & Stanton, 2017a, 2017b, 2017c; Walch et al., 2015).

2.4   COMMUNICATION DURING AUTOMATED DRIVING

When attempting to address the emergent issues of level 3 and 4 automation outlined in Section 2.1.2, it is important to address them in their entirety. As is with any complex engineering system, making changes may improve one outcome but degrade another. Many individual studies approach the handover with limited hypotheses in mind but few test interaction designs in their entirety. The following section provides an overview of distributed situation awareness (DSA), a leading situation awareness theory for complex sociotechnical systems and the theory base of 'joint activity' (JA)—the concept of applying human–human communication principles to human–automation interaction. Together DSA and JA can bring insights into ensuring that C/HAVs collaborate effectively and ensure that transactions are optimized, safety focused, and address a broad range of interaction outcomes.

2.5 DISTRIBUTED SITUATION AWARENESS

Reductions in SA is well cited as a contributing factor toward incidents in many domains (e.g., Endsley & Kiris, 1995; Gold et al., 2016; Horswill & McKenna, 2004; Jentsch et al., 1999; Stanton et al., 1997), including those with automation capabilities. SA has changed a lot since its inception; the original description of SA proposed by Endsley (1995) states that SA represents the accurate perception, comprehension, and projection of situational elements. However, many researchers and practitioners are now favoring a distributed cognition approach to the concept of SA—DSA (e.g., Salmon et al., 2009, 2016; Sorensen & Stanton, 2016; Stanton et al., 2006, 2017b). This is due to the recognition of complex sociotechnical systems consisting of both human and nonhuman agents, each interacting with different perceptions and interpretations of the environment. From this perspective, it is the entire system that either gains or loses SA (Salmon et al., 2016). This approach is beneficial as it acknowledges variations in individual 'schemata'—the cognitive templates built over time informed by experience (Neisser, 1976; Stanton et al., 2006)—and role expectations that each agent brings to the task and demonstrates how information does not have to reside equally among every individual, as this information can be accessed when required in environmental artifacts, a concept that traditional models do not address.

Stanton et al.'s (2006, 2017b) theory of DSA moves situation awareness toward one that encompasses a distributed cognition framework. DSA proposes that each agent in a system has a different interpretation of the situation based on previous experience, different information available to them, and differing types of 'awareness' about what is going on due to the cognitive and physical constructs of the agent (see Stanton et al., 2017b, 461). Situation awareness, therefore, emerges through 'transactions' (Sorensen & Stanton, 2016) between individual agents, as well as interacting with the environment and individually held schemata (Neisser, 1976; Stanton et al., 2006).

For C/HAV handover, regaining control from an automated system may lead to degraded driver performance as a result of deskilling or incompatible SA (Sorensen & Stanton, 2016; Stanton et al., 2006, 2017b; Stanton & Marsden, 1996). While automation is active, the driver is separated from the driving task, thus emphasis should be placed on the interactions made between automation and driver to ensure system SA is raised during vulnerable periods such as the handover of control.

To apply the insights generated by DSA, Salmon et al. (2009) provide a number of recommendations when implementing DSA theory to a given system. For C/HAV handover, the following are deemed to be particularly pertinent and worthy of further discussion:

- SA requirements should be clearly specified as a result of careful analysis.
- C/HAVs should be designed to support SA transactions.

- Unwanted information should be removed.
- Interfaces should be customizable.
- C/HAVs should provide appropriate and explicit communication links.
- C/HAVs should use procedures to facilitate DSA.

Throughout the book, these recommendations are addressed in relation to improving SA in C/HAV interaction for design and practice. From a DSA perspective, viewing the handover as the only time information can be exchanged between driver and vehicle is unrealistic, information can continue to be relayed through transactions between both driver and vehicle. DSA specifications therefore require the driver to be able to access information they require, when they require it.

This approach, therefore, supports a need-to-know information transfer at the time of handover and allows for access to information through interfaces following the handover. Further, instilling a sense of training and standardized communication between driver and automation allows for a system to be optimized with regards to raising system SA and becoming more efficient. Interfaces can be improved by being customizable (and thereby more appropriate to the user and context), by delivering more accurate information, and by providing information that can be understood by the driver. This is a unique perspective, as a human and a machine will inevitably have a different sense of the environment as they process information differently (Stanton et al., 2017b).

For C/HAVs, the DSA approach may be particularly beneficial for the following reasons: 1) drivers and automation will inevitably hold different perspectives on the environment both due to previous activities and unique capabilities. 2) C/HAVs will be able to process, feedback, and feed-forward (i.e., what is happening and what will happen) and potentially take control of aspects of the driving task (e.g., automatic braking) and direct the driver to hazards (e.g., visual and audio alerts) during all stages of the driving task, whether in primary control or otherwise. 3) Information is provided to the driver not as a way of 'sharing SA' but matched to the driver's unique capabilities and role, fostering what is known as 'compatible SA' through 'transactions' (Sorensen & Stanton, 2016). This may be a more appropriate approach to vehicle automation, as differences between humans and computers result in the requirement for information to be presented in line with a driver's cognitive abilities (Salmon et al., 2009).

2.6 JOINT ACTIVITY

As a way of addressing DSA and the role that communication has in human–machine teams, the theoretical framework of 'joint activity' (Clark, 1996; Klein et al., 2004, 2005; Bradshaw et al., 2009) may help guide C/HAV designers and manufacturers toward creating a system that facilitates collaborative interactions while addressing the distributed nature of SA.

Coordination within tasks is a central concept in human–human communication (Clark, 1996), computer-mediated communication (Monk et al., 2003), and human–agent interaction (Bradshaw et al., 2009). Klein et al. (2004) discuss the steps required to make automation a 'team player', which has led to further discussions surrounding coordinative concepts in human–agent interaction (e.g., Bradshaw et al., 2009; Klein et al., 2005). As noted by these authors, the coordinative concepts that make up JA provides designers of automation with a clear understanding of how to alleviate breakdowns in communication. JA themes include:

- Agreement to collaborate—Both agents intend to work toward common goals.
- Mutual predictability—Both agents are clear with their intentions and can reliably predict their counterpart's future actions.
- Directability—Instructions or advice regarding the situation.
- Goal management—Mutual goals are established.
- Common ground—Mutual understanding and common knowledge are confirmed.
- Communicate capacity—Agents communicate their ability to perform tasks.
- Signal phases—Agents indicate what phase their role in the task is in.
- Coordination devices—Agents should use common artifacts to guide collaboration.

In the context of C/HAVs, JA suggests that a collaborative interface would involve an interaction that allows agents to register whether the other agent is capable of performing, what they are processing, gauging what their future intentions might be, and being able to direct them toward next steps. Further, examples of goals could include destination or how long automation should be used for. This could be pre-planned prior to or during the journey.

JA has been selected among other communication theories as it directly addresses the issues outlined by Sarter et al. (1997), who state that automation that remains silent can lead to limits and capacities being exceeded without the human operator's knowledge. JA features human–human communication roots and sets itself apart from other communication theories due to its development within the automation domain and provides practical steps toward ensuring that automation surprises and breakdowns in communication do not occur (Bradshaw et al., 2009; Klein et al., 2004, 2005). Other such theories of communication, such as uncertainty reduction theory (the theory in which relationships develop as predictability increases; Berger & Calabrese, 1974), or symbolic interactionism theory (the process in which communication occurs through meaning, language and thought; Hewitt & Shulman, 1979) may contribute toward aspects of human–automation interaction; however, these theories do not directly task communication and focus on a wider range of aspects within communication such as identity and cultural influences. These theories provide cultural, social, and more abstract interpretations of the design issue, whereas JA is useful for the context of C/HAVs due to its development for application in human and machine teamwork domains.

2.7 SUMMARIZING THEORIES

Both DSA (Stanton et al., 2006) and JA (Clark, 1996; Klein et al., 2004, 2005) provide valuable insights into how interactions can be performed. JA promotes 'a willingness to invest energy and accommodate to others, rather than just performing alone in one's narrow scope and sub-goals' (Klein et al., 2005, 94). Together with the DSA philosophy that relates to presenting the right information at the right time to the right team member, a mixed theoretical approach can identify useful and need-to-know information but implement it in a way as to not overload the receiver with derelict information that may not be of use. Both DSA and JA promote the philosophy of the 'economy of effort', minimizing costs incurred via communication while ensuring that each agent has the required information at the appropriate time.

DSA and JA have been selected together, as each framework addresses the short-comings of the other. For example, DSA does not provide detailed guidance on how interactions should occur on a microlevel, rather, DSA outlines major considerations that should be made to optimize roles and transmission of information. On the other hand, JA provides more detail as to how both humans and automations should communicate and outlines a variety of central themes that should be adhered to. JA's major drawback is that it does not consider how each agent has a unique interpretation of the environment, and therefore, it will inherently have different requirements from a situation awareness perspective.

This book proposes that both JA and DSA can be used to identify what, when, and how information should be relayed between both driver and automation across a variety of contexts in C/HAVs. Their advantage over competing theories within the automation domain is that the tenets outlined by JA and DSA are readily attuned for interface design within human–machine systems as they directly address how and what information should be transferred to ensure safe operation. It follows that for the design of interfaces and interactions, both JA and DSA are well equipped for this book's target outcomes. Other such theories that have proven influential for improving safety in human–machine teams, such as the Contextual Control Model (CoCoM; Hollnagel, 1993), focus on elements of teamwork but fall short of informing what and how information should be communicated between agents. The CoCoM, for example, focuses on strategy and temporal context to understand whether a situation is under control or whether agents are merely reacting to environmental cues. It is a useful tool to categorize and avoid control breakdowns and thereby increase safety; however, both JA and DSA provide insights that not only aim to ensure safe operation but also calibrate trust, workload, usability, and acceptance.

With DSA and JA together, the proposed approach outlined in this book for dealing with issues within C/HAV interaction is to view automation as a copilot rather than a tool, a move that a number of researchers support (e.g., Klein et al., 2005; Eriksson & Stanton, 2017c; Stanton, 2015). The capabilities for agents to collaborate, coordinate, and execute tasks in line with mutual goals and expectations is likely to have a great impact on a number of factors including safety, efficiency, and trust.

This integration of DSA and JA guides the chapters of this book. Chapters refer, where appropriate, to the concepts outlined in this section to provide a guideline for C/HAV interaction design.

2.8   FUTURE DIRECTIONS

To begin the process of designing a novel C/HAV interaction design, the next chapter draws on the theoretical concepts identified in this chapter to map the domain of C/HAVs using cognitive work analysis (CWA; Vicente, 1999). In doing so, future work in the field will help us understand how values identified as being capable of improving communication can be implemented into designs. This can be achieved by illustrating the physical objects, system goals, and tasks that must be filled by both driver and automation.

2.8 FUTURE DIRECTIONS

3 Cognitive Work Analysis to Improve Communication in AV Interactions

3.1 WHAT IS COGNITIVE WORK ANALYSIS?

Bridging the gap between theory and practice can be challenging when starting the design process. Stanton et al. (2017b) recommend mapping the sociotechnical domain prior to beginning investigations to gain a better understanding of the processes at work. This chapter draws on the theoretical concepts outlined in Chapter 2 to improve communication in conditionally and highly automated vehicles (C/HAVs) and applies them using cognitive work analysis (CWA; Vicente, 1999), a method capable of identifying constraints and affordances present in a given system to target improvements toward a predefined value. CWA is an effective tool for mapping out complex sociotechnical domains and exploring the roles of agents within a system and is comprised of multiple steps. These steps can be selected and combined based on the questions asked and the unique domain under analysis. In doing so, an overview of how communication can be improved in C/HAV interaction can be developed. CWA was selected as it directly addresses actors within a given domain during the social organization and cooperation analysis—contextual activity template (SOCA-CAT). The SOCA-CAT can provide valuable insights into how agents with different roles can achieve the domain's values and objectives (Vicente, 1999). In this way, it addresses the issue of function allocation, an aspect that is of great importance within the field of automation (Fuld, 2000; Idris et al., 2016).

CWA has been applied to automotive issues as a way of tackling design problems with a specific outcome (Allison & Stanton, 2018; Birrell et al., 2012; Salmon et al., 2007; Stanton & Allison, 2020). For human–machine interface (HMI) designers, the use of CWA can outline the system's capabilities/constraints to ensure that communications can be optimized while ensuring the breakdowns and errors are prevented. It allows for an informed approach on how the user can engage with such technology and can be created considering the values/priorities of the system. For C/HAV interaction design, CWA will provide insights into how, what, and when automation and driver should communicate.

The stages of CWA include:

- Work domain analysis—A global exploration of processes, objects, and values in a domain and how they are connected.

DOI: 10.1201/9781003213963-4

- Control task analysis—An analysis of the tasks that are required to take place to ensure success within the domain.
- Strategies analysis—An analysis of what strategies can be implemented for a task along with selection criteria including time pressure, difficulty, and risk level.
- Social organization and cooperation analysis—The distribution of tasks in the domain to agents.
- Worker competencies analysis—To factor in knowledge, rules, and skills of workers to ensure that capabilities are not exceeded and to provide a structure for defining training and system design requirements.

For this analysis, work domain analysis, control task analysis, and social organization/cooperation analysis were selected as these steps map the overall domain, the tasks required to take place, and the allocation of function within these tasks. These stages are most appropriate to address the research questions in this book as they allow for the identification of HMI elements, map out the stages of automation, identify tasks to perform to ensure effective communication, and assign them to agents in line with the concept of function allocation (Fuld, 2000; Idris et al., 2016).

3.1.1 WORK DOMAIN ANALYSIS

Work domain analysis consists of an abstraction hierarchy (AH)—a tool that aims to capture an image of a system, its components, and provide insight into how they can be applied to overarching values using a five-tier structure to represent a system independent of specific tasks or activities (Jenkins et al., 2008). A series of means-ends connections filter through the tiers of the system through the tiers down to the physical components of the system. A given node indicates the 'what', the connected nodes above indicate 'why' the selected node is present, and the connected nodes below illustrate 'how' the node can be implemented.

To begin, physical components are identified separately to achieve the functional purpose. The analyst then identifies how the functional purpose can be measured and specifies these measurements as system values. Next, they summarize the physical components (object-related processes) into what they can contribute toward system function (e.g., instrument cluster, steering wheel, and accelerator pedal). The step of connecting the top two and bottom two tiers involves identifying purpose-related functions that show how physical components can contribute to the overall desired purpose of the system. This allows analysts to understand how designs can target certain aspects of the system to achieve the overarching design goal.

3.1.2 CONTEXTUAL ACTIVITY TEMPLATE

Leading on from work domain analysis, the CAT allows analysts to take purpose-related functions or object-related functions and assess the situations in which these functions can be but typically is not applied (dashed lines; e.g., the driver could be alerted to an upcoming handover) and can be applied while typically being the case (bow-tie plots; e.g., due to safety criticality the driver must be alerted to the handover when it is impending). These figures are plotted against situations or locations that may affect the nature of the work being carried out.

3.1.3 Social and Organization Cooperation Analysis

Function allocation is a vital and recurring theme throughout automation literature (Fuld, 2000; Idris et al., 2016). Assigning roles to either human or automation correctly ensures that the most optimal performance can be achieved from the system. The SOCA can be used for the purpose of understanding the roles that both automation and driver can be assigned to during the automated cycle. This stage extends the CAT so that each stage and each process can be described in relation to which agent is acting at that stage.

3.2 DEVELOPMENT OF ANALYSIS

3.2.1 Participants

The analysis was developed by the primary researcher (academic researcher, 3 years of HF experience, male, aged 26). A subject matter expert (SME), an automotive engineer working directly with automation and behavior (male, aged 46, 12 years of experience in leading design projects in ADAS/autonomous features), verified this analysis.

3.2.2 Abstraction Hierarchy

3.2.2.1 Identifying Functional Purpose
The functional purpose for the CWA was drawn from the research outcome for this book outlined in Chapter 1, Section 1.3.1.1, summarized as: 'facilitate effective communication between driver and automation'.

3.2.2.2 Identifying Design Values/Priority Measures
The theory from distributed situation awareness (DSA) and joint activity (JA) can help inform communication concepts in automated vehicle (AV) operation. To that end, values were drawn from both DSA and JA with the intention of creating a handover interface that promotes safe and efficient communication. The additional themes added by DSA and JA are outlined in Table 3.1. These themes seemed appropriate and overarching to the outcomes desired through a handover assistant that ensures that the goals of both DSA and JA are addressed in harmony with one another.

3.2.3 Identifying Physical Components

Physical components were generated by assessing the available tools to manufacturers through assessing current level 2 and 3 AV manuals which included Tesla S-Class (Tesla, 2020) and Audi's A8 (Audi, 2019b).

TABLE 3.1
Values/Priority Measures Derived from Theory for Application to Abstraction Hierarchy

Value/Priority Measure	Contributory Theory
Maximize distributed situation awareness	DSA
Optimize calibration of trust	JA
Maximize coordinated activity	JA
Maximize usability	DSA/JA
Maximize efficiency	DSA
Maximize safety	DSA/JA

3.2.4 IDENTIFYING PURPOSES AND FUNCTIONS

Object-related purposes link physical objects that perform a similar task. For example, a head-up display and an instrument cluster both provide the driver with visual information. Purpose-related functions connect object-related purposes with priority measures to bring together the work domain analysis, summarizing each process in a way that relates to the goals of the analysis. These links were made through researcher deliberation, checking manuals where necessary, and confirming with a SME upon completion.

3.2.5 CONTEXTUAL ACTIVITY TEMPLATE/SOCIAL AND ORGANIZATION COOPERATION ANALYSIS

For the SOCA, purpose-related functions were drawn from the AH and considered against the stages of semiautomated handover. These stages were derived from a number of sources including McCall et al.'s (2016) taxonomy of the notification and the event, for both driver-to-vehicle handback and vehicle-to-driver handover. Pre-journey situation was derived from the ability of drivers to interact with their devices prior to their journey (e.g., for trip planning and customization purposes). Information stages were added due to the requirement for SA transactions to take place, much like that of shift handover (Adamson et al., 1999; Cohen et al., 2012). Finally, the post-handover phase was appended as automation still has capabilities even when control is primarily with the driver, this also mirrors that of shift handover in air traffic control where shifts stay for a set amount of time to supervise the incoming operator (Federal Aviation Administration, 2010; Kontogiannis & Malakis, 2013; Walker et al., 2010).

3.2.6 DEVELOPMENT OF ANALYSIS

One full-day meeting was held with the SME to verify the output generated from the analysis. The SME was briefed on the theory and the rationale behind the analysis. Following this, the SME provided their insights into the physical components available in AVs and provided their viewpoints on the node connections that were made in the WDA and how the functions and processes were generated as a result of the analysis. Next, the SME provided input to the tasks that were to be carried out during the process of automated driving and how actors may contribute to each stage.

3.3 RESULTS

3.3.1 WORK DOMAIN ANALYSIS

Figures 3.1–3.3 present visual re-renderings of the AH to show nodes that address driver–automation interaction. For the AH, the *functional purpose* for addressing automation at this stage is to 'optimize communication between driver and handover assistant'. This specification allows a breadth of factors to be addressed in the *values and priority measures*.

As a synthesis of both DSA and JA, four values represent the measurable outcomes that should be targeted to achieve the functional purpose. The AH includes 'maximize safety', 'maximize efficiency', 'maximize clarity', 'increase coordinated activity', 'optimize calibration of trust', and 'maximize distributed situation awareness'. Safety, efficiency, and maximizing situation awareness are related to DSA, which outlines that transactions should take place between driver and automation to ensure that operation can remain safe after control transitions have been performed. Further, DSA suggests that information should be tailored as to not be overly saturated with the aim to increase system SA but optimize performance—any knowledge that can remain potentially accessible and within automation should be identified. Clarity and coordinated activity relate to JA. Establishing an understanding between both driver and automation increases clarity about the situation and ensuring that each agent communicates with the other about changes to their own goals and expected outcomes of the journey and the tasks involved ensures that both automation and driver are coordinated with their actions. Uncalibrated trust can lead to misuse of a given system indicating that trust should align with system capability (Lee & See, 2004; Walker et al., 2016).

The level below outlines *purpose-related functions* that connect the artifacts available within the system with the values that the system aims to address. In this level, nodes such as 'facilitate bidirectional communication' and 'communicate function status' address the aforementioned values by promoting interactions between driver and automation with specific actions in mind. Below the center tier are the *object-related processes* that summarize the individual physical objects and artifacts available into how they could contribute to the overall analysis. In this analysis, nodes such as 'provides visual information to the driver' (e.g., connected to head-down display, head-up display, and center console) and 'facilitates trip planning' summarize all the physical objects that could allow the driver to interact with the vehicle with the purpose of customizing the functions of the AV. Finally, the physical objects make up the bottom tier identifying components within the system that could have a contribution toward the analysis.

Following the links down from the upper tiers outlines 'how' that node can be addressed, whereas following links up from a lower tier illustrates 'why' that node exists and what it can contribute toward. With a complete AH, other analyses can be addressed using the outcomes generated from the AH. The following sections outline the contributions that various HMI elements can contribute to the overall communication process between driver and vehicle. The figures displayed in the following relate to HMI elements. The full AH addresses the entire system and shows how each physical object maps to the overall processes, functions, and measures of a level 3/4 AV.

3.3.1.1 Visual Displays

The physical objects within Figure 3.1 represent what can be visually provided to the driver during the automation cycle. Following the links upward, visual modalities address every value and priority measure specified in this AH. These interfaces together can display relevant pieces of information related to safety, state, and capability. Of particular interest is the role of the center console. These devices double up as an input device as well as a visual display (visual information can be displayed and interacted with through touch-screen capabilities). This adaptive functionality may allow drivers to plan trips, input customizations, and have a record of current progress. This object can be utilized during automation to make amendments to plans and receive instant feedback regarding the situation. Nomadic devices in this analysis indicates a device, such as a tablet, which is exterior to the vehicle's inbuilt system. This could be a paired device or one supplied by the manufacturer to allow for both infotainment and communication (e.g., to display an alert).

3.3.2 Vocal and Audio Communication

Vocal communication can add an additional control link between driver and vehicle. Figure 3.2 illustrates the CWA framework for this particular modality. This two-way

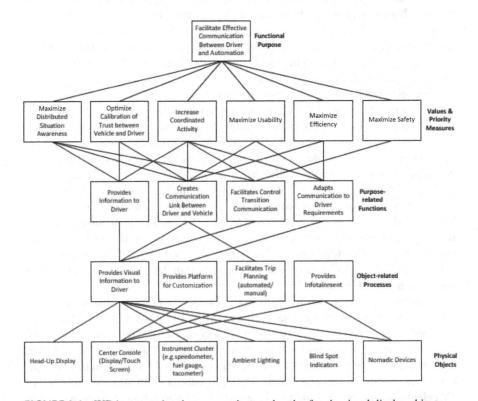

FIGURE 3.1   WDA output showing connections and nodes for six visual display objects.

process allows the driver to provide commands or requests to the AV with feedback being presented either visually or through audio. Tracing the 'why' paths from microphone upward, it is evident that inputs can allow the system to assess the awareness of the driver, as well as serve as a suitable way of addressing the full range of values and priority measures outlined in the AH. Vocal communication with audio feedback addresses similar factors as visual displays showing that it is a suitable way of communicating the concepts outlined by the JA and DSA.

3.3.3 PHYSICAL INPUTS

Physical inputs in the form of buttons, scroll wheels, paddles, and dials can provide the driver with instant communication in a one-way fashion. Inputs allow drivers to change settings, potentially during a journey, so that both driver and vehicle can collaborate as the driving task is performed. The analysis showed that inputs have three processes to allow communication, to allow customization, and to facilitate trip planning. In doing so, these inputs address every value and priority measure outlined for effective communication during the automated cycle. Figure 3.3 illustrates the CWA framework for physical inputs.

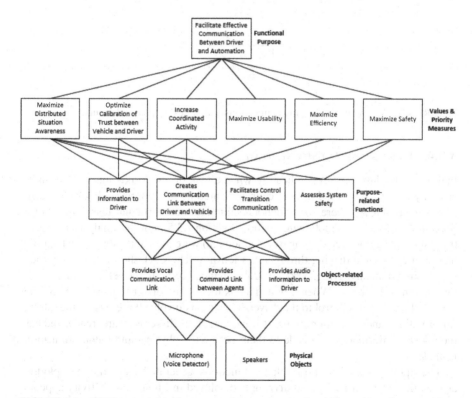

FIGURE 3.2   WDA output showing connections and nodes for audio and vocal communication objects.

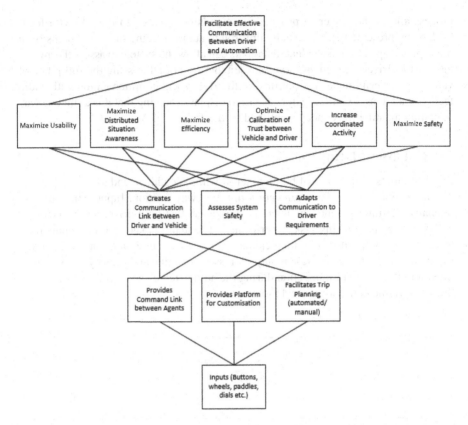

FIGURE 3.3   WDA output showing connections and nodes for physical inputs.

3.3.4 CONTEXTUAL ACTIVITY TEMPLATE

Figure 3.4 displays the CAT indicating when a task on the Y-axis must occur, indicated by bow-tie plots, and might occur, indicated by dashed boxes. Figure 3.4 also displays the SOCA, where colors denote actors for each particular task. This is outlined in detail within the following section (Section 3.3.5). For this analysis, the identified stages of automated driving (see Section 2.2) were used to illustrate the dynamic nature of automated driving throughout a journey. The CAT shows the pre-journey leading into manual driving. Next, control is 'handed back' to the vehicle and it enters into automated driving. When a boundary is met or the driver requests control, the vehicle 'hands over' control to the driver. Transitions involve three stages, the 'alert', 'information', and 'transition' stages, to ensure that the driver is aware, ready, and has the relevant information available to them before the physical and mental transition is made.

The application of purpose-related functions generated from the AH plotted against the stages of automated driving is displayed in Figure 3.4. Activity appears to cluster around transition periods for the purpose of communication; however, it is important to note many processes can continue to take place at every stage of the automated cycle. Many of the processes can take place during the pre-journey, such as collaborating goals, agents directing one another toward important tasks that are required to be handled, and communicating future intentions. A great deal of customization can be done at this stage to facilitate the specific requirements of the journey prior to manual driving.

3.3.5 SOCIAL ORGANIZATION AND COOPERATION ANALYSIS

Figure 3.4 is broken down into subcomponents to better illustrate which element of the automation is required to act during the automated cycle. For processes such as 'maintain safe control of the vehicle', it is notable that primary responsibility is with either driver or automation based on the availability and status of automated control. This no longer becomes true during transitions, as at these times safe operation is shared, as either agent making an error could threaten the safe control of the vehicle. For processes that involve vehicle-to-driver communication, this is naturally the role of the automation takes. For others that require both agents to contribute to the communication process, shared action is indicated.

The overall contribution of a communication system is illustrated in green. Multiple paths are possible to communicate between driver and vehicle as well as align mutual goals and collaborate during the automation cycle. Aside from haptic information, this communication link has the opportunity to take place throughout the entire automated cycle. Sensory systems should be available throughout the entire cycle, with detection being implemented mandatorily when automation is responsible for vehicle control.

Figure 3.4 serves as a helpful tool in diagnosing and identifying what aspects of the C/HAV system is required to act during each stage of the automation cycle. The SOCA-CAT identifies the importance for multiple automation systems to be considered and shows how it fits in to the overall performance of the system. The analysis can serve as a guide for which systems exist in C/HAVs as well as the stages that must be performed during a journey. This template will be used for designing each stage of the automation cycle within the design sections of this book and will also serve as a useful discussion point in which designs can be referred to in both the design and testing stages of the book.

3.4   DISCUSSION

This chapter has provided an analysis of components, processes, and function allocation for level 3/4 AVs through the implementation of cognitive work analysis (Vicente, 1999). The nature of automated driving is inherently complex. The roadway environment involves a multitude of factors such as weather, visibility, condition of roads, road layout, other road users, and pedestrians (Paxion et al., 2014). Creating a single communication standard to apply to these conditions will be a challenge; however, understanding the components available to designers and how they can address the issue of JA can serve as a foundation for the development of novel prototypes.

Where other domains have addressed both automation and the transfer of control (e.g., Bainbridge, 1983; Idris et al., 2016, Sheridan, 2002), the AV domain is targeted

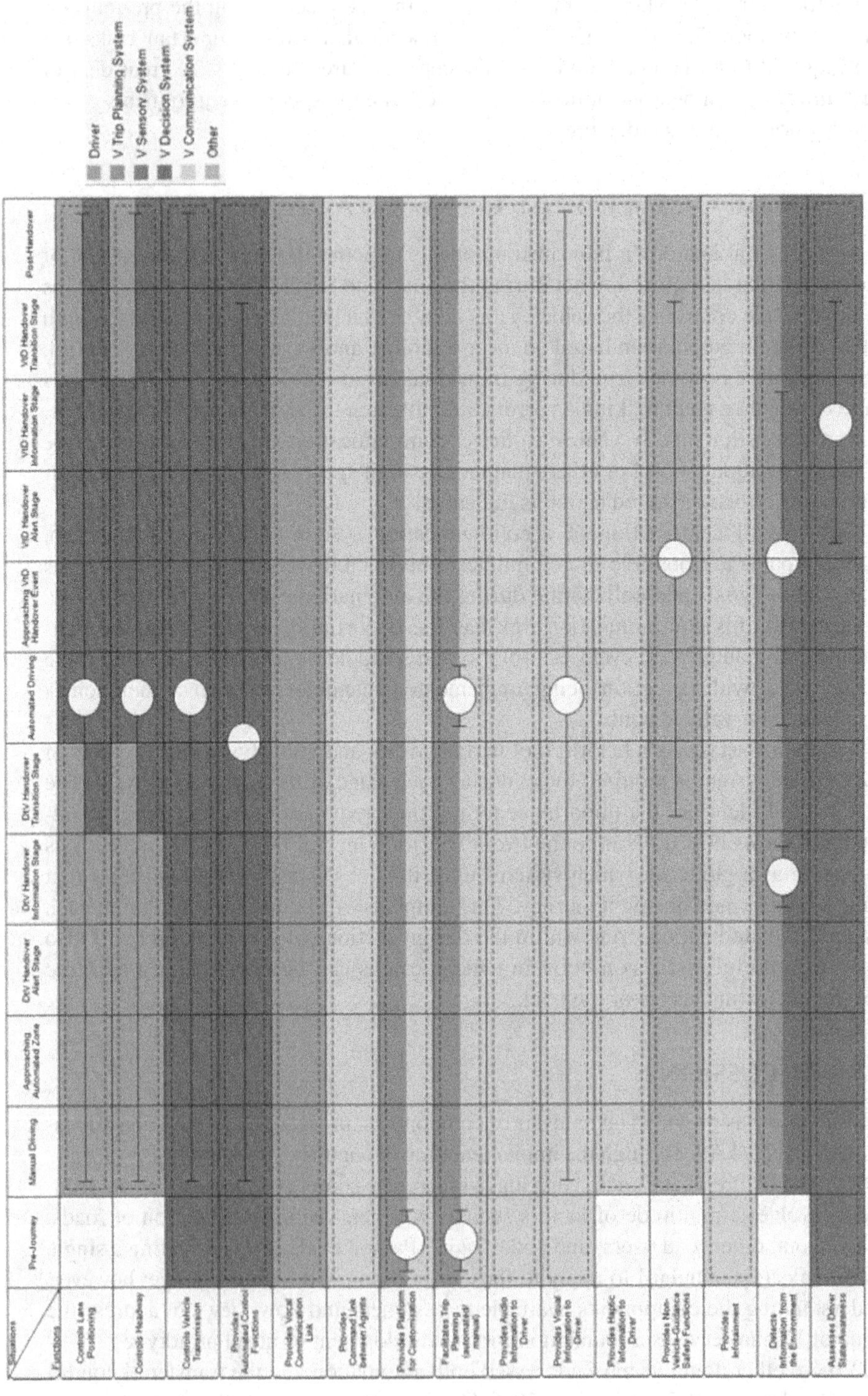

FIGURE 3.4 CAT and SOCA-CAT for optimizing communication in level 3 and 4 AVs.

primarily at the general public. This factor is noteworthy, as emergent issues involve whether the standards of training meet safety requirements, whether users will purchase the vehicle, whether the driver feels that automation is working with them for a common benefit, and indeed, whether automated system is trustworthy for activation in certain areas. Interaction with AVs, therefore, must address effective collaboration in JA, the optimization of trust, usability, and ensure that the system's distributed situation awareness does not degrade to unsafe levels during the transfer of control (Clark, 1996; Klein et al., 2004, 2005; Salas et al., 2000; Sheridan, 2002; Sorensen & Stanton, 2016). This analysis aimed to converge knowledge on shared control to directly account for these factors by implementing CWA, with the goal of providing a roadmap for designers and manufacturers alike to understand what is available to them during concept generation. As a result, domain constraints have been identified to allow designers and manufacturers to identify physical mechanisms in place and denote each of them a role by the AV system (including the driver) to achieve optimal communication during the automation cycle.

The key findings from this analysis are as follows:

- There are many methods for communicating between vehicle and driver. Among these methods, there is great variety in visual modalities available to designers; however, physical inputs and vocal interaction may facilitate a two-way process for the driver to initiate commands.
- The automated cycle can be broken down into 12 stages with two directions, each direction including the following stages: control, approach, alert, information, and transition.
- The domain can be attributed to 15 purpose-related functions (see Figure 3.4).
- The domain can be mapped by denoting five roles: the driver, trip-planning system, sensory system, decision system, and communication system.
- Control responsibility is devoted to either driver or automation depending on whether automation is active; however, an overlap occurs during the transition stage and the period following the vehicle-to-driver handover.

From this analysis, it is clear that there are many options available to AV designers to create a clear and effective route for communicating between automation and driver. Designers interested in the process should consider which systems should be responsible for transmitting safety information (see Clark et al., 2019a for insights into information transmission during AV handover) and which systems can be used to communicate secondary information such as trip planning. In doing so, collaborative communication can be achieved and in turn reduce degradation in system SA when transfers of control are performed (Endsley & Kiris, 1995; Stanton et al., 2017b).

When navigating the output of this analysis, it is important to understand what the situation is and whether it requires situational-based considerations—notably, the direction of control, which party initiated the exchange, and whether the scenario is that of an emergency or a planned handover (Mirnig et al., 2017). These are all important aspects to consider when deciding on which modalities should convey information to and from the driver. Modalities that are situated closer to the road environment (such as head-up display) may be more suitable when the driver's attention should be paid toward the road (Clark et al., 2019c), whereas updates or information where driver control is not imminent could be displayed on the center console. Additionally, notifications from audio/vocal feedback could be present to allow for quicker delivery of information—research has showed that vocal communication is a suitable way of directing a driver's visual attention toward important areas of interest, in turn, addressing the role of directability found in the JA framework (Clark et al., 2019a, 2019c).

Observing the AH shows that inputs are a means of allowing the system to assess safety, perhaps through prompting, questioning, or indeed analogous to the so-called dead-man's switch where inputs indicate that the driver is conscious and potentially aware of their surroundings. When the transfer of control is present, these factors are of utmost importance, as control transitions when the receiving agent is not prepared can lead to great vulnerabilities (Endsley & Kiris, 1995). In turn, the bidirectional command link between automation and driver allows each agent to communicate (in advance or real time) their intentions, observations, and capabilities in line with JA and increasing system situation awareness through transactions (Klein et al., 2004, 2005; Sorensen & Stanton, 2016). Some modalities are instantaneous, whereas others may require additional processing time or additional factors such as the ability to attract attention. These factors should be considered while selecting physical inputs and/or voice interaction to allow the driver to send messages to the vehicle. These input features are also central to trip planning and setting customizations. However, literature suggests that more information is not always useful. Overloading drivers by utilizing everything available may require attentional resources in excess of those available. When testing prototypes, a suitable workload, trust, and usability analysis (e.g., NASA Task Load Index, Hart & Staveland, 1988; System Usability Scale, Brooke, 1996; and System Acceptance Scale, Van der Laan et al., 1997) should be performed to ensure that drivers engage with the system intuitively, safely, and in a user-friendly fashion.

This analysis proposes the automation cycle stages outlined in Figure 3.4 with the intention of allocating functions to different agents. The CAT draws on object-related functions to define what needs to be and what can be performed during these stages. Unsurprisingly, control functions must be performed throughout—although this role is allocated primarily to driver or automation depending on the stage of the automation cycle. Notably, when control is transferred between agents, an overlap occurs that may persist into the next stage. An overlap may involve a gradual control transfer or shared responsibility, with the possibility of providing the incoming agent with supervision—a potential mitigation to degraded SA, as illustrated by shift-handover practices (Clark et al., 2019b). For the purpose of effective communication, the SOCA-CAT shows that four automation agents can be allocated to object-related processes in a plan-sense-decide-communicate fashion. Aside from the decision system, these agents are assigned a role that does not vary over the automation cycle. Notably, detecting information from the environment can be done by either driver or automation-sensor system throughout the cycle.

It is noteworthy that this analysis, as with many other human factor's analyses, is susceptible to being biased toward the researcher's interpretation of the research question and domain. Further, although SMEs serve as a useful confirmation for work carried out by academic researchers, this analysis would be improved by including additional experts from a wider range of demographics.

Going forward from this analysis requires thorough consideration into what is required to optimize communication during the automation cycle. This analysis provides six values that may help measure the effectiveness of communication: situation awareness, calibrated trust, usability, coordinated activity, safety and efficiency. By measuring these variables in prototype tests, a researcher can gauge how well their communication system is performing. These values can be addressed by utilizing objects and their associated processes through the appropriate agent. Prototypes may be formed through a combination of physical components that address a wide range of purposes and functions, and only together, will they address these values sufficiently.

3.4.1 FUTURE DIRECTIONS

Chapters 2 and 3 provide the foundations for designing a communicative HMI in C/HAVs. This initial exploration of the nature of communication within C/HAVs will be applied throughout the remaining chapters of this book. The CWA featured in this chapter will be applied to the piloting stage through the identification of which HMI elements are available for testing (e.g., head-up display, cluster, and speakers) and applied to the design stage to set the scene for participants to design around (e.g., the stages of the automation cycle and the modalities that are available; Chapter 7). Finally, the CWA will inform design discussions in Chapter 8 with regards to what needs to happen throughout the cycle of automation, giving the design stage a firm grounding in both theoretical and practical aspects of C/HAV interaction.

To further understand how situation awareness can be raised to counteract vulnerabilities in task handover, human teamwork domains can provide insights into the communicative strategies required to preserve safe task continuation when safe operation is critical. Chapter 4 represents the final chapter in the scoping stage (Clark et al., 2019b). It provides an in-depth overview of shift handover in domains such as medicine, aviation, and energy manufacturing to provide current and future domains requiring the continuation of activity with a range of effective strategies that adhere to the principles of distributed situation awareness, before implementing select strategies to C/HAV handover.

4 Review of Handover Tools and Techniques in High-Risk Shift-Work Domains

4.1 INTRODUCTION

So far, Chapter 2 has outlined the theoretical foundations for improving communication between driver and automation in conditionally and highly automated vehicles (C/HAVs) and Chapter 3 has developed on this by linking theory tenets with the domain through cognitive work analysis (CWA). How solutions can be implemented is becoming clearer—a combination of visual and vocal interaction ensures that domain values are met. However, the details surrounding handover protocol, the continuation of the working task, and what should be communicated during critical events remain unidentified.

As mentioned in Chapter 3, it is tempting for researchers to replicate the aviation domain when implementing novel applications of human–automation interaction. Domains such as commercial airlines feature shared-control features and transitions much like that of C/HAVs, leading this domain to face many of the same issues as that of C/HAVs (see Salmon et al., 2016). The domain is undoubtedly influential; however, there are several factors that researchers may overlook that make the automated vehicle (AV) domain unique, suggesting that a range of strategies should be considered and tested in C/HAV interaction. There are several factors that make the C/HAV particularly unique:

- AV operation may not have procedures in place for effective planning— Prior to takeoff, pilots rigorously plan route activities (Federal Aviation Administration, 2017). In C/HAVs, drivers may wish to alter goals more dynamically in line with the driver's plans for the journey.
- Users are the general public—Commercial pilots who engage in automated activities are those who have typically been recruited by the organization and take part in more rigorous training (Federal Aviation Administration, 2020). As driving is more inclusive and available to the general public, users may also be representative of a broader range of backgrounds and experiences.
- Automation is marketed as a quality-of-life service—Aviation typically utilizes it for 'safe and efficient operation' by allowing pilots to direct attentional resources to other tasks. In driving it is currently marketed as

DOI: 10.1201/9781003213963-5

contributing to safe, comfortable, efficient, and enjoyable personal travel (Khan et al., 2012; Stanton et al., 2001; Ward, 2000), whereby drivers can release time to take part in secondary activities (e.g., Fagnant & Kockelman, 2015) and become more accessible to those who may otherwise not be able to manually drive (Alessandrini et al., 2015). Current automation systems can be activated and deactivated in line with user desires.

- Fewer opportunities for training—Whereas 1,500 h are required to be a fully qualified commercial pilot with an airline (FAA, 2020), currently, the UK government states that there is no minimum number of lessons required to pass driving tests (GOV, 2012).

Domains to be learned from are not limited to human–automation domains. Wherever continuous tasks and human operators are required, effective communication is required to ensure the task maintains high performance. This chapter builds on the current work explored in the previous chapters by considering a range of strategies currently in use across major high-risk shift-work domains. They are then summarized using the theory of distributed situation awareness (DSA). These strategies intend to provide a platform to inform the design of C/HAV interaction and other domains that may require novel techniques for raising situation awareness (SA) during handovers.

4.1.1 The Handover of Control and Responsibility

In work environments requiring continuous human-to-human activity (for example, air traffic control (ATC), healthcare, military, maritime, and energy generation and distribution), logistical boundaries dictate the need for a handover of control and responsibility between personnel (Stanton et al., 2010). Boundaries may include high levels of operator fatigue, an imminent breach of operational capabilities, or the requirement of a specialist. The handover task creates issues for safety, as incidents disproportionally occur directly following handover (e.g., Thomas et al., 2013). These incidents are typically attributed to inadequacies in the transfer of 'situation awareness' (Stanton et al., 2017b) during the handover period. The challenge that many domains must overcome is how outgoing personnel can effectively encourage compatible SA for incoming personnel and foster a seamless and safe handover of control and responsibility. This review identifies and summarizes the handover tools and techniques (HTTs) that high-risk domains involving human–human handover use to achieve better communication during shift change and discusses these HTTs through the lens of DSA theory (Stanton et al., 2006). This is with the intention for current practitioners to evaluate their own, potentially new, domain and identify which HTTs may be suitable for their domain's unique requirements.

Throughout the literature, many terms have been used to represent the handover process and the variety of steps involved. The terms identified by the reviewers were handover, handoff, takeover, sign out, shift change, shift-to-shift report, transition of care, exchange of control, and position relief briefing (Eurocontrol, 2007; Fassert & Bezzina, 2007; Federal Aviation Administration, 2010; Riesenberg, 2012; U.S. Department of Transportation, 1995; Wilkinson & Lardner, 2013). Many of these definitions have their own applications to the specific domain in which they are used (e.g., transition of care in healthcare settings). Given that all of these terms relate to the process of an incoming agent taking responsibility/control from an outgoing agent, we use the term 'handover' to apply to the collection of these terms.

A number of authors have attempted to create structure for application to the handover task. Grusenmeyer's (1995) phases of shift change outline that the handover occurs over four stages: the end of shift, the arrival of the incoming operator, the meeting, and the taking of post. McCall et al. (2016) have provided a simpler interpretation of two phases: the 'notification' and the 'event'. One common misunderstanding is the distinction between 'handover' and 'takeover', as both have been used interchangeably in the research literature (Walch et al., 2015). Recent reviews have clarified that 'takeover' refers to the moment of the incoming party regaining control and the outgoing party relinquishing control (Eriksson & Stanton, 2017b; Merat & de Waard, 2014; Morgan et al., 2016). Following this framework, we define the 'handover' as the entire process beginning with the 'notification' from either party and 'takeover' as the moment that control is relinquished. As a result of this review, optional steps during the handover are also presented in Figure 4.1.

Domains typically have a protocol when it comes to the handover. Many make use of unique HTTs to guide the handover task. For the purpose of this review, the reviewers define HTTs as encompassing communication strategies, handover aids, and any other action/method adopted to attempt to improve communication during the handover process.

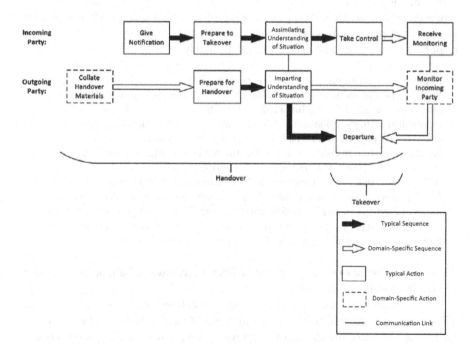

FIGURE 4.1 Flowchart of generalized actions taking place during the shift-handover process.

4.1.2 APPLYING DISTRIBUTED SITUATION AWARENESS TO THE HANDOVER TASK

During its inception and its emergence in the literature, SA has had a central focus on the individual—the human in which a task is concerned (Endsley, 1995; Stanton et al., 2006, 2017b). The most influential SA model outlines perception, comprehension, and projection as being the constructs that make up SA. As systems become more complex, this approach is becoming limited in its scope, as it does fail to address how information is stored, distributed, and can interact with its networked components (Stanton et al., 2006, 2017b).

Due to the complexity of the handover task and the numerous artifacts involved, SA should be addressed via a systems approach and view the handover task as a collection of transactions in SA between components within a system (Stanton et al., 2006, 2017b). DSA suggests that to ensure that a system works effectively, a system must acknowledge that each individual and component (e.g., interfaces, sensors, and automation) has its own perception of the situation and for humans, a unique understanding of the situation viewed through their own experience. Therefore, rather than merely share information about the situation, SA should be made compatible between components through means of transactions related to the task and roles in which individuals experience (Stanton et al., 2006, 2017b). DSA implements a cyclical approach inspired by the perceptual cycle model (Neisser, 1976; consisting of world, perception, and schemata), where the system has overall SA that is dynamically changing in line with environmental cues—perhaps as a result of actions taken by the components within the system (Salas et al., 1992; Stanton et al., 2017b). Schemata, as defined by Neisser (1976), comprises both genotype and phenotype schemata. These are described as consisting of schemata that are already present as a result of previous experiences and schemata that are dynamically created as a response to the activity and interactions, respectively.

Regarding the handover task, there has been much discussion related to raising SA. Studies typically address the perception of situation elements (e.g., de Carvalho et al., 2012; Durso et al., 2007; Le Bris et al., 2012). This feature of SA has been discussed in practical settings in the form of 'information transfer' (IT), which relates to the effective sharing of information between groups of individuals within and between organizations (see Borowitz et al., 2008; Bulfone et al., 2012; Stanton et al., 2017b; van Wijk et al., 2008). The DSA approach describes IT as providing the receiver with 'transactions' that can be integrated with their own schemata (Sorensen & Stanton, 2016). Further, transactions are bidirectional as both the receiver and deliverer become aware of each other's awareness (Salmon et al., 2009; Sorensen & Stanton, 2016). This view focuses on each agent building their own SA for application to their own particular tasks and goals while relating to their own experience and training.

As a result of work exploring the role of DSA in teamwork, Salmon et al. (2009) outline 16 guidelines (see Table 4.1).

For an in-depth discussion regarding these guidelines, see Salmon et al. (2009). These guidelines provide practitioners with a way of improving system performance, where need-to-know information is displayed appropriately for that individual and their role. How well these guidelines can be applied to current handover practice is

TABLE 4.1

Distributed Situation Awareness Design Guidelines

Guideline No.	Guideline
1	Clearly define and specify SA requirements
2	Ensure roles and responsibilities are clearly defined
3	Design to support compatible SA requirements
4	Design to support SA transactions
5	Group information based on links between information elements in DSA requirements analysis
6	Support meta-SA through training, procedures, and displays
7	Remove unwanted information
8	Use customizable/tailored interfaces
9	Use multiple interlinked systems for multiple roles and goals
10	Consider the technological capability available and its impact on SA
11	Ensure that the information presented to users is accurate at all times
12	Ensure information is presented to users in a timely fashion and that the timeliness of key information is represented
13	Provide appropriate and explicit communication links
14	More information is not exclusively better
15	Use filtering functions
16	Present SA-related information in an appropriate format

yet to be determined, a recurring issue cited across the literature is 'how much is too much?' with regards to information exchange (transactions).

4.1.3 Purpose of the Review

Past reviews in handover protocol have been limited in scope, either by interpreting strategies for their application to a specific field of practice (Lardner, 2006; Lawrence et al., 2008; Morgan et al., 2016; Patterson et al., 2004; Plocher et al., 2011; Riesenberg, 2012; Thomas et al., 2013), focusing on a narrow set of studies (Patterson et al., 2004; Raduma-Tomas et al., 2011), exploring the domains rather than comparing and contrasting the protocols implemented (Wilkinson & Lardner, 2012), or becoming outdated (Patterson et al., 2004). Further, no review has yet applied the theory base of DSA (Stanton et al., 2006) to the handover task. This review aims to collate, compare, and contrast literature regarding the handover tools/techniques (HTTs) used in a variety of domains during handover, to discuss them in light of DSA, and to summarize them based on the design recommendations made by Salmon et al. (2009).

4.2 METHOD

4.2.1 Search Methods and Source Selection

Many key terms such as 'handover', 'handoff', and 'shift change' relate to radio technology, chemistry, biology, and physics. Where possible, search terms were filtered to home in on relevant handover literature. Sources were searched for with the terms displayed in Table 4.2 using Web of Science, Google Scholar, and Scopus. The titles of the first 1,000 results from each search were reviewed, ordered by relevance on Google Scholar, citation count on Web of Science, and all results from Scopus to ensure that as many key papers were captured as possible. Subsequent search terms were then adapted to target-specific domains based on keywords found in previously identified articles.

Many industries report their protocol in large organizational reports; therefore, a wealth of information regarding an industrial practice can be gleaned from a handful of these reports. To supplement search, the bibliographies of four key papers and reviews of major domains were reviewed and selected for their breadth of knowledge and their relevance to hard-to-access areas (i.e., Catchpole et al., 2007; Plocher et al., 2011; Lardner, 1996; Patterson & Woods, 2001).

In total, 799 sources were identified. The vast majority of sources were related to the healthcare domain (698, 87.36%), followed by aviation (40, 5%), energy manufacturing (35, 4.38%), and other domains such as military and railroads (7, 0.87%). The remainder of sources were classified as being unaffiliated (19, 2.37%). The representation of handover in the domain of healthcare is likely due to the importance of patient safety, resulting in higher amounts of research attention and funding.

TABLE 4.2
Syntax Used during Literature Search

Syntax	Search Tool	Result Count	Action	Unique Identified Articles
Intitle: handover(s) OR 'hand over(s)' OR handoff(s) OR 'hand off(s)' OR signout(s) OR 'sign out(s)' OR signover(s)	Web of Science (filtering out telecommunication-related fields)	1,884	Titles of top 1,000 most cited reviewed	528
OR 'sign over(s)' NOT radio(s) NOT network(s) NOT mobile(s) NOT wireless AND LANGUAGE:(English)	Google Scholar	~17,200	Titles of top 1000 most relevant reviewed	134
	Scopus (filtering out telecommunication-related fields)	378	All titles reviewed	96
Intopic: 'shift handover'	Web of Science	45	All titles reviewed	9
Intopic: 'shift change' OR 'shift changes' OR 'shift changeover'	Web of Science Filtered for relevant handover domains	129	All titles reviewed	9
Intopic: handover AND oil	Web of Science	30	All titles reviewed	4
	Identified from key papers	—	All titles reviewed	19

Consequently, in some domains, the handover process is more likely to be mentioned in papers or books related to human cooperation or the causes of specific accidents rather than having articles or books dedicated to the issue.

All 799 titles, and where necessary, abstracts, were reviewed on their contribution to the handover strategy literature in their relevant domain either by discussing or proposing handover strategies to be implemented. Around 419 sources met this criterion. Of these, 376 papers were related to healthcare handover. To create a platform for equal representation across domains, only medical sources that mentioned handover 'strategies' in their abstracts were included in the final review. Of these, 40 medical sources were carried over to the review, alongside the 43 from the other domains totaling 83 sources. The final sources were comprised of a range of source types: 28 literature reviews, 19 quantitative studies, 16 qualitative studies, 11 organizational documents, 4 mixed methods documents, 4 design papers, and 1 discussion paper.

4.3 RESULTS AND DISCUSSION

4.3.1 OVERVIEW OF HANDOVER TOOLS/TECHNIQUES

In total, 19 HTTs were identified within the literature. Table 4.3 outlines the frequency at which HTTs were discussed, in each domain, as a viable tool in aiding the handover of responsibility.

Commonalities include the demand for standardization, vocal communication, and making use of technology during handover. However, domains appear to differ in many aspects regarding the handover procedure. The healthcare domain focuses on training programs and the use of contextual information (e.g., Anderson et al., 2015; Haig et al., 2006; Iedema et al., 2009; Bost et al., 2012), whereas energy manufacturing focuses on the accumulation and review of accurate past information (e.g., Adamson et al., 1999). Aviation pays a particular emphasis on clarifying control and overlapping responsibility through monitoring the operation of their counterpart (e.g., Federal Aviation Administration, 2010).

During this review, it was found that most HTTs, in some way, have been discussed in an empirical framework, whether that be through interpretations of case studies or measurements during controlled trials. The most cited HTTs have been assessed using both objective and subjective approaches. Other HTTs, such as 'clarify control' and 'contextual handover', appear to have been developed in response to domain constraints or vulnerabilities. Finally, a minority that are poorly represented in the literature (e.g., guided walkthrough) so appear to have no explicit findings to show that they are effective. It is important for domains to validate their approaches to their handover tasks to ensure that they are effective for their specific situations.

4.3.2 STANDARDIZING HANDOVER PROTOCOL

With two thirds of the sources including some discussion around the standardization of handover protocol, it can be safely concluded that the majority of research attention has been paid toward establishing a domain-wide approach to handover protocol.

TABLE 4.3

Authors' Count of Sources Discussing Each HTT As a Viable Method

No.	HTT	Healthcare	Aviation	Energy	Military	Maritime	Unaffiliated	Total
	[Total number of sources]	40	20	17	1	1	4	83
1	Standardization	31	13	7		1	2	54
2	Vocal communication	16	9	9	1	1	2	38
3	Use of past information	12	6	14	1	1	1	35
4	Training programs	19	3	5			2	29
5	Bidirectional exchange	13	7	8			1	29
6	Use of technology	12	7	4	1		1	25
7	Face-to-face	10	3	6	1		2	22
8	Adaptation	11	3	4			1	19
9	Compatible mental model	7	6	4	1		1	19
10	Preparation	5	3	7	1		2	18
11	Contextual handover	13	1	1			1	16
12	Read-back	8	1				1	10
13	Clarify control	2	7				1	10
14	Third parties	7	1				1	8
15	Overlap of vigilance	1	6	1			1	8
16	Assess handover	3		2				5
17	Shared responsibility	4	1					5
18	Multiple media			3			1	4
19	Walkthrough		1					1

Note: HTTs are ordered by total number of mentions as a viable HTT across all domains

Many authors from across the domains agree that a standardized handover protocol reduces the likelihood of critical information being omitted (e.g., Adamson et al., 1999; Brazier & Pacitti, 2008; Dawson et al., 2013; Norris et al., 2014; Riesenberg et al., 2009a) while also ensuring that critical information is not subject to any bias or misinterpretation (Gross et al., 2016). By far the most favored strategy toward standardization is the adaptation of a structured checklist or mnemonic to the specific domain/setting in which it is to be applied. Forty-seven of the fifty-two identified sources discuss their application.

It is of no surprise that the content of checklists is vastly different between domains. Domains differ on the type of information as well as the content, for example, Patterson et al. (2004) note that the healthcare domain cannot assess patient status 'at a glance' and requires a holistic view on the patient's condition, whereas ATC operators can take advantage of transmitting information in a predictable fashion (e.g., air pressure will always be required and either be referred to as high/low/min stack; Fassert & Bezzina, 2007).

An example of the most discussed case of standardization comes from the medical domain, with 18 sources discussing SBAR (situation, background, assessment, and recommendation) as a viable structure for handover. SBAR was designed to ensure the transmission of a mental model, as well as reduce cognitive demands during handover meetings (Arora et al., 2008, 2014; Cheung et al., 2010; Haig et al.,

2006; Riesenberg et al., 2009b). SBAR addresses SA sufficiently, as it gives the individual a sense of the previous events and the rationale behind actions to allow for them to perform the task effectively following the handover (Haig et al., 2006).

Previous studies indicate that structuring information during handover has the capability to reduce absolute medical errors made (see Horwitz et al., 2007; Starmer et al., 2014a, 2014b). These structures address SA by presenting relevant information at the perceived correct timing (Salmon et al., 2009). Many of the structures found for the clinical domain within this review focus on a holistic interpretive account of the situation, whereas those used in ATC appear to be more descriptive. An example comes from the National Air Traffic Services (NATS) in the UK who make use of mnemonics such as PRAWNS in the approach environment, outlined in the following (Walker et al., 2010; Wilkinson & Lardner, 2012):

- P—Pressure (barometric)
- R—Runway currently in use
- A—Area sector information and how they are organized
- W—Weather conditions
- N—Nonstandard Priority Information
- S—Strip data for aircraft status

The mnemonics within ATC typically include information about the current situation that will affect future decisions without explicitly transmitting past information or current goals but rely on the operator's decision-making and schemata to interpret this information. The content of these mnemonics is specific and does not involve the same encompassment of a mental model as checklists such as SBAR. However, air traffic controllers make use of other HTTs to supplement handover, such as monitoring radars prior to and after handover, which will be discussed in the relevant section.

Energy production and distribution control room operators favor longer and more detailed checklists to ensure that physical checks have been complete prior to takeover (Lardner, 2006). In this domain, leaving out one detail can lead to disastrous consequences. Examples include that of the Buncefield fire incident in 2005 caused in part by a miscommunication of which pipeline was filling which tank. This has since been attributed to many deficiencies in shift changeover protocol including a lack of handover structure resulting in uncertainty around whether key information had been transmitted; in doing so, operators collectively lost SA in relation to the current operational status (Brazier & Pacitti, 2008; Gordon, 1998; Wilkinson & Lardner, 2012, 2013; Stanton et al., 2010). To avoid incidents such as Buncefield, checklists are used to cover risk factors related to handover and ensure that all information related to previous, current, and future work has been included (Adamson et al., 1999; Department of the Army, 2007; Lardner, 2006; Wilkinson & Lardner, 2012, 2013).

From a DSA perspective, checklists allow domains to identify 'need to-know' information and ensure that it is delivered at the appropriate time to prevent unwanted information from being addressed and to reduce inefficient information exchange. However, problems may arise during the implementation stage. For example, SBAR is applied to a range of arguably contrasting circumstances across healthcare. A rigid checklist that is generally applied loses the specification factor as it may not be tailored to that individual role/task. Providing shift-workers with standardized approaches to the handover tasks may ensure that schema, and SA, is compatible among agents prior to the handover task. Further, it may allow for the domain to implement what they believe is necessary information during this time leading to an optimization in the amount of information transferred (Sorensen & Stanton, 2016; Stanton et al., 2006).

As a final note on this HTT, there is a warning to be made regarding strict overreliance on standardization. Cohen et al. (2012) contest the use of checklists by stating that the implementation of a rigid checklist can result in a bias toward one-way transmission of information. Further, if a checklist is to be used, it is important that the checklist be tailored for the specific domain and sub-domain in question (Staggers & Blaz, 2013)—a poorly constructed checklist can result in longer handover times and increase the amount of irrelevant content transmitted—in line with the DSA guidelines put forward by Salmon et al. (2009).

4.3.3 Vocal Communication, Face-to-Face and Bidirectional Exchange of Information (HTTs 2, 5, and 7)

Throughout the literature, HTTs 2, 5, and 7 were discussed either on their own or as a collective. This section will discuss them together as they all relate to exchanges made from one operator to another.

Written information has been described as being insufficient for handover when used as the singular stream of IT (Adamson et al., 1999; Brazier & Pacitti, 2008; de Carvalho et al., 2012). Research over recent years has focused on the importance of the vocal exchange of information during handover. Sources typically distinguish between written and verbal handover, where verbal relates to the use of voice rather than text. These sources have been classified as 'vocal' in our review for clarity. In nursing, it has been suggested that rather than replacing the current flexibility of handover with 'over-rigid standardization', vocal communication skills can be developed to enhance the handover and is applicable to all nursing situations (Randell et al., 2011). Further, in space mission-control, it is common for operators to refuse to take control unless a vocal update has been given (Patterson & Woods, 2001).

Research in healthcare settings suggests that vocal communication directs attention toward priority information more readily than written information (Chui & Stone, 2012). Other suggestions from healthcare, aviation, and energy domains suggest that vocal communication provides a platform for feedback on how well information is being received (Adamson et al., 1999; Federal Aviation Administration, 2010; Parke et al., 2010; Parke & Mishkin, 2005; Walker et al., 2010). From a DSA perspective, face-to-face and vocal communication provides an explicit avenue of communication which allows for immediate two-way feedback to be given and ensures that the shift taking over can receive the information that they require through the use of questioning. Such advantages should be taken advantage of, as this provides domains with the support required for transactions in SA to take place.

The healthcare domain mentions face-to-face interaction more so than other domains, likely due to the enhanced capability of the environment. Other domains are more likely to conduct tasks at a workstation, thereby hindering face-to-face interaction. Hobbs (2008) notes that in aviation maintenance face-to-face interaction can be used to transmit nonverbal communication to gain additional information using gestures and emphasis (Philibert, 2009). Further, face-to-face interaction is less effortful and immediate (Lebie et al., 1995).

Research also suggests that one-way information transmission is not as effective as two-way interaction (Cohen et al., 2012). The role of questioning has been a major focus of handover literature, placing a requirement for the incoming staff member to engage in cooperative interaction to facilitate the handover procedure (Drach-Zahavy & Hadid, 2015; Parke & Kanki, 2008; Rayo et al., 2014). As an example, Parke and Mishkin (2005) illustrate that during NASA's Mars Exploration Rover (MER) mission, featuring the rovers 'Spirit' and 'Opportunity', three major handovers took place. These involved lengthy face-to face meetings involving two-way interactions so that questions could be asked. Through asking questions, gaps in the knowledge of the incoming operator can be filled, and rather than relying on the transfer of a descriptive account, the receiver can create their own mental model (Bost et al., 2012; Revell & Stanton, 2012). This is done through the access of additional information that may not have been transmitted originally.

Errors and incidents may occur when a culture of questioning is not permitted (Sutcliffe et al., 2004; Wachter & Shojania, 2004). Many organizations have written this into their official protocol (e.g., Eurocontrol, 2012; Lardner, 2006; The Royal College of Surgeons of England, 2007) noting that the handover should be a two-way process, giving the person about to take responsibility an opportunity to ask questions.

4.3.4 Use of Past Information

In ATC, space-shuttle operations, and energy manufacturing control rooms, knowledge of what has happened is suggested to be important in the process of raising SA in individuals, as this allows operators to understand current and future operation and goals (Adamson et al., 1999; Kontogiannis & Malakis, 2013; Patterson & Woods, 2001; Stanton et al., 2017b). Twelve of the fourteen (86%) energy manufacturing domain sources reviewed in this chapter discussed reviewing logs, making this their top priority during the handover.

In energy manufacturing and distribution, the handover of correct and accurate information is conducted to avoid scenarios such as the major Buncefield explosion mentioned in HTT 1 (see Wilkinson & Lardner, 2013). Another example of this is the Piper Alpha explosion (Lardner, 1996; Paté-Cornell, 1993) that has been partly attributed to the failure to transmit information about a removed safety release valve for a condensate pump during shift changeover. The next shift encountered an issue with a second condensate pump and made the decision to restart the (unbeknownst to them) compromised pump, giving way to the resulting explosions that occurred shortly after. Knowledge of such consequences may be why more experienced operators are more likely to check previous trends and information when coming onto shift (Li et al., 2011). However, relying heavily on personalized logs and notes has been criticized as being understructured and should include structures to ensure priority information is transmitted (Plocher et al., 2011). If this HTT is to be adopted, thought should also be paid to the structure and the layout of handover logs, alongside trials to assess their effectiveness.

The DSA approach favors such an HTT, as it allows for information to reside in the system, without relying heavily on one-to-one communication (Stanton et al., 2006). However, this approach does not provide an immediate explicit communication, although it may serve as a safety net should the incoming shift require information without having to establish communications with the previous shift.

4.3.5 TRAINING PROGRAMS

Errors that occur during handover may be due to insufficient training (Li et al., 2012). When implementing a structured handover, providing individuals with the appropriate training has been reported to be an effective HTT (Gordon & Findley, 2011; van Sluisveld et al., 2013; Pucher et al., 2015; Weikert & Johansson, 1999). This has been achieved in a number of ways including giving guidance on an implemented structured tool, enhancing communication skills, building trust between staff members, and taking part in simulations of handover scenarios (Drach-Zahavy & Hadid, 2015; Gordon & Findley, 2011; Horwitz et al., 2012; Pucher et al., 2015). Under the DSA guidelines providing such training programs may allow for individual schema to be addressed and ensure that agents are compatible during their transactions (Neisser, 1976; Stanton et al., 2006).

Many organizations across domains note the importance of training and practice on handover effectiveness (Eurocontrol, 2007; Fassert & Bezzina, 2007; Patterson et al., 2004; Weikert & Johansson, 1999). An example of a program from the healthcare domain is the 'HELiCS program'. This program makes use of video playback of real-time scenarios so that personnel can develop their handover communication through discussion and in-depth analysis (Bost et al., 2012; Iedema et al., 2009).

4.3.6 USE OF TECHNOLOGY

The DSA theory base places an emphasis on the use of technology and its role in SA. Information residing in displays, sensors, and automation is no longer supplementary to humans but rather forms an element of the system as a whole (Stanton et al., 2006). In current day operations, technology plays a central role in team communication. Bolstad et al. (2003) explored the ways in which SA can be raised during army operations. Their review recommended aspects such as video conferencing, file sharing, networked radios, and program sharing to exchange information. All reviewed domains make use of technology in the form of either electronic handover tools, electronic logs/health records, or video data to handover more effectively (Cheung et al., 2010; Hannaford et al., 2013; Parke & Kanki, 2008).

Literature has repeatedly outlined the importance of designing human–machine interfaces with the human operator in mind (Hopkin, 1989; Stanton, 1993). In ATC, this has been particularly important as the development of technologies such as sensing devices, computer assistance, and prediction services have changed the landscape of how humans interact with their work setting over time (see Nolan, 2010). These interfaces can be optimized to foster a smoother handover (Brazier & Pacitti, 2008; Hopkin, 1989). Today's air traffic controllers use radars and electronic flight strips at their workstations to raise SA during the handover period (Durso et al., 2007; Kontogiannis & Malakis, 2013; Walker et al., 2010).

Lawrence et al. (2008) discussed technological possibilities for coordinating handover in a chaotic emergency department involving color changes of screens 2 h before handover and a blinking 40 min prior. The use of electronic time tracking allows staff to preempt the handover with enough time to prepare. Further, studies have explored the benefits of using electronic systems rather than paper systems to handover shifts, and they report improvements to the continuity of care, likely due to the increased accessibility of information (Cheah et al., 2005; Raptis et al., 2009). Allowing for information to be distributed across the system in computer, as well as human agents, ensures that the system keeps high SA, without requiring the transmission of unnecessary information during handover (Stanton et al., 2006).

4.3.7 ADAPTATION OF TASK OR SETTING

Situational factors, such as the timing and location of the handover, may affect handover. The location of handover is deemed as being important due to distractions such as noise or staff interruptions posing a threat to the effective transmission of information (Cheung et al., 2010; Spooner et al., 2015). In healthcare settings, the location of handover significantly varies between institutions (Street et al., 2011). A call to standardize the location for handover has been made over the years, as this ensures access to data systems, be away from distractions, and allow confidential information to be passed on (Chui & Stone, 2012; Singer & Dean, 2006). Douglas et al. (2017) also discuss multitasking in healthcare and include the handover as one of their considerations. They draw upon van Rensen et al.'s (2012) findings that conducting vocal handover after monitoring equipment had been prepared was more effective and no more time consuming than doing both concurrently.

The timing of handovers is also important so that individuals are well prepared for the handover (Eurocontrol, 2007). In domains such as energy manufacturing, the incoming operator must be aware of the upcoming handover so that no attention-critical events are managed during this time. Setting timetables for the handover allows personnel to plan time effectively for tasks such as checking records. Further, this allows them to make an early arrival to ensure handover goes smoothly (Wilkinson & Lardner, 2012, 2013). A tactical example of adaptation comes from ATC. By conducting the handover during a low workload period, the deterioration of control can be avoided (Durso et al., 2007; Walker et al., 2010). ATC tasks are shifted toward achieving short-term goals while putting requests on hold to ensure that the incoming operator can handle the issues in their own way (Durso et al., 2007).

4.3.8 COMPATIBLE MENTAL MODEL

Being able to synchronize goals and establish a narrative has long been regarded as the goal of the handover. This is facilitated through understanding how humans process information (Grusenmeyer, 1995). A large focus in some of the identified sources is specifically related to how raising SA is not done through receiving information alone but rather being able to relate information to schemata, tasks, and temporal features of the environment (Stanton et al., 2017b). A mental model is defined as a mental representation of how the real world operates and can be applied to a given task (Revell & Stanton, 2012). Every individual possesses a mental model regardless of whether it is accurate or not (Revell & Stanton, 2012), in line with the DSA approach of compatible SA between agents (Stanton et al., 2017b). Therefore, by assuming that the outgoing operator's mental model of the situation is correct, the goal of the handover is to transfer SA and ensure that the incoming party has an adequate mental model of the situation.

With the popularity of SBAR in healthcare settings, the focus on a shared mental model has been successful in the healthcare domain. Cohen et al. (2012) outline that individuals have mental models as a summary of the information available to them so they can pass on to the following shift. They describe how similar mental models during handover allow for slight differences to approaches that may not have been considered previously. LeBaron et al.'s (2016) recent exploratory analysis describes how intensive care unit (ICU) physicians coordinate their actions by communicating a mental image of 'where we were, where we are, and where we're going' (p. 520).

4.3.9 Preparation

Setting time aside to handover is important to ensure that materials are in place, although in areas such as energy distribution, this appears to be commonplace (Stanton & Ashleigh, 2000). This is likely due to less pressure on time limits and more detailed information to read through (Adamson et al., 1999; Wilkinson & Lardner, 2012, 2013). A period for preparing and reviewing handover material was also utilized in the MER handover procedure (Parke & Mishkin, 2005).

4.3.10 A Contextual Handover

Our review found that domains differ greatly on the likelihood that a specific set of information will be required for handover. Domains such as ATC involve predictable information types, whereas in healthcare settings the need to adapt to different patient statuses, needs, and requirements requires a flexible handover procedure. Due to many sub-domains in healthcare settings being present, some researchers have made it clear that a domain-wide standardization may not be possible (Anderson et al., 2015). If handover is overly structured, details regarding the patient's unique condition may be omitted and the capability for medical staff to make up for this through use of previous experience may have an impact on the quality of the handover (Staggers & Blaz, 2013).

Patterson (2008) warns against the unintended consequences of standardizing a domain-wide rigid checklist. In reality, trade-offs have to be made due to external pressures, and staff must use their intuition by deviating from the template. Consequently, they may be criticized by authorities for any failure that may occur due to their deviation, even if they are fully justified in their deviation due to the contextual aspects involved. Therefore, HTTs that are resilient to these environmental pressures should be favored (Drach-Zahavy et al., 2015), as well as HTTs that take into account a range of local factors (Bulfone et al., 2012). As an example, the energy production domain favors allocating more time to handover if the incoming staff have been absent for a longer period of time (Adamson et al., 1999). This approach better ensures that SA is raised to the required level based on the current state of the incoming shift.

4.3.11   OTHER HANDOVER TOOLS AND TECHNIQUES (HTTs 12–19)

A number of additional HTTs were also identified. These HTTs have been grouped into two distinct groups: those related to handover techniques and those concerned with handover quality. The handover techniques are read-back, clarification of control, use of multiple media, and walk-throughs. Those concerning handover quality are the presence of a third party, overlap of vigilance, assessments of handover quality, and shared responsibility. Each following paragraph in this session outlines a given HTT.

4.3.11.1.   Additional Techniques

- The use of read-back involves the receiver repeating back to the sender information that they receive. This approach can be used as a way of ensuring the accurate transmission of information and of correcting any errors. This HTT may have an additional benefit, reviewed by Macleod et al. (2010), who refer to the improvement of memory simply by saying a word aloud. This technique has shown to improve the handover procedure in the healthcare domain (Boyd et al., 2014; Brown, 2004; Patterson et al., 2004).
- In a plane's cockpit, who is in control of the aircraft can be unclear at times. In these scenarios, verifying who has control during handover can be useful. Some crews state 'you have control' with the person taking control instantly replying 'I have control' to ensure that both parties are aware of the transition of control (U.S. Department of Transportation, 1995).
- Arora et al. (2008, 2014) describe how a culture of shared responsibility can help manage the negative effects of the handover. By working as a team with shared goals and mental models, handover can be made more effective compared to doctors treating their patients as 'their own'. ATC also practices this, as both the outgoing and the incoming operators are tasked with the effective transmission of information (Federal Aviation Administration, 2010).
- The energy manufacturing domain makes use of guided walk-throughs during handover to ensure that incoming operators have seen the status of the facility. This is conducted alongside a vocal talk through and allows the outgoing operator to remind himself or herself of all information required for handover and allow incoming operators to see firsthand the status of their environment (de Carvalho et al., 2012).

4.3.11.2. Handover Quality

- Many sectors find that conducting the handover with the presence or monitoring of a patient or an authoritative figure can have a positive influence on IT. Particularly, errors can be corrected and logistics can be better managed (Flink et al., 2012; Patterson et al., 2004; Tobiano et al., 2013).
- In ATC, it is common for operators to monitor one another's task handling prior and after the takeover (Federal Aviation Administration, 2010; Kontogiannis & Malakis, 2013; Walker et al., 2010). During the arrival stage of the handover, the incoming operator scans the radar to familiarize themselves with the airspace and the strategies taken by the current operator. They achieve this by plugging in his or her headset into a communications port to listen to outgoing and incoming transmissions. After the takeover of control, the outgoing operator oversees the new operator for a brief period to ensure the tasks are being dealt with appropriately (Durso et al., 2007; Walker et al., 2010).
- Many sources have noted that finding methods to measure the quality of handover is important to ensure that current HTTs and factors related to staffing are adequate (Brazier & Pacitti, 2008; Lardner, 2006).
- Multiple media are also used to provide information in a variety of ways, which help to encompass a range of information types. Sources claim that supplying handover information vocally and written information concurrently is a viable method for improving handover communication (Brazier & Pacitti, 2008; Lardner, 1996).

4.3.12   HTTs AND THE DSA GUIDELINES

Regarding the HTTs generated in this review, many have been identified as being of importance within DSA research (e.g., training programs, standardized protocol, and assessing performance; Salmon et al., 2009). From this review, there are notable insights that could be drawn upon. HTTs that focus on the availability of accurate information for accurate representations of the situation (e.g., standardization, technology, and contextual handover) can be applied to the DSA approach of transactions of information regarding the situation when, and where, it is necessary (Stanton et al., 2006, 2017b). For example, training programs can be utilized to foster suitable schemata for better comprehension of these cues and address an individual's schema, in line with the perceptual cycle model and DSA (Banks et al., 2018; Neisser, 1976; Stanton et al., 2006).

Techniques such as overlap of vigilance rely on individual SA representations within each operating pair, so that if SA is not compatible (perhaps in the form of missing cues or not comprehending/projecting the scenario sufficiently), then the outgoing operator has the opportunity to correct this if they detect a danger in the current operation. Additionally, standardizing information to understand what is 'need-to-know' may also provide practitioners with a way to reduce the amount of unwanted information present and allow for an efficient delivery of information. By extension, following protocol may also allow individuals to know what is expected of them, who knows what, and who needs to know what and when. This can be aided by the use of technology through addressing a number of guidelines related to the presentation of information and storing information within electronic sources.

HTTs such as bidirectional exchange and implementing a 'contextual handover' may allow for shifts to tailor information to their own individual needs/roles. By doing so, unwanted information can be filtered, allowing incoming shifts to retrieve only the information that is required.

Domains have a variety of human factors analysis techniques (such as task analysis; Annett, 2003, or CWA; Vicente, 1999) to refine their protocol, checklists, and training programs to facilitate transactions and improve system performance. It would no doubt be beneficial to a domain to implement a combination of HTTs into a handover protocol to address all the recommendations made by Salmon et al. (2009) for raising DSA.

4.4 CONCLUSION

Nineteen HTTs were identified in a range of high-risk domains such as healthcare, aviation, and energy manufacturing/distribution. Domains differ in their approaches to the handover procedure, although many similarities exist. Popular HTTs include the use and adaptation of checklists, two-way interaction with questioning, and the use of past information. This review has provided a discussion of these HTTs in relation to DSA and provides a unique perspective on the handover task. Many of the HTTs identified address DSA in a variety of ways—we propose that to maximize benefits, multiple HTTs should be adopted so that guidelines presented by Salmon et al. (2009) can be addressed in their entirety.

4.4.1 FUTURE DIRECTIONS

This chapter considers shift handover separately from C/HAV handover to give a thorough overview of how handover could occur. An understanding of how strategies can be applied to C/HAV handover is still required. Chapter 5 discusses how human-team communication may contribute to knowledge of how handover assistants should be designed (Clark et al., 2019b). Vocal strategies are tested in a driving simulator by replicating a shift-handover task specifically for the driving task.

Section II

Pilot Testing These Concepts in Automated Driving

5 Replicating Human–Human Communication in a Vehicle

A Simulation Study

5.1 INTRODUCTION

Chapter 4 outlines a wide variety of tools and techniques designed to raise situation awareness in human teams within a system where handovers are required (Clark et al., 2019b). To begin to apply the strategies to C/HAV handover this chapter explores potential vocal communication strategies by replicating human–human handover in a C/HAV simulation. Due to humans being capable of dealing with advanced requests, it is assumed that outcomes of this study will help inform driver–automation interaction for handover assistants. This assumption is discussed in the following section.

5.1.1 APPLYING HUMAN COMMUNICATION THEORY TO HUMAN–MACHINE HANDOVER

There has been much debate over how applicable human–human communication (HHC)/computer-mediated communication (CMC) is to applications of human–computer interaction (HCI). Several communication researchers suggest that the nature of communication for joint activity is similar, regardless of the agents present (e.g., coordination, shared goals, and resolve breakdowns in communication; Klein et al., 2005). This line of thinking has also been discussed in the field of automation (Klein et al., 2004). Bradshaw et al. (2003) second this standpoint by showing that joint activity via HHC or HCI involve what they call 'collaborative autonomy', that is, the process of understanding, problem solving, and executing certain tasks.

Perhaps the largest body of work regarding similarities between HHC and HCI comes from Reeves and Nass (1996), who outline their 'Media Equation' that illustrates how humans interact with technology as if they are real people. For example, it has been shown that factors such as cooperation, aggression, courtesy, and trust could play a role in these interactions. Although the original work may be outdated by today's technological standards, the research inspired a wave of investigations regarding how technology can interact with cultural and social mechanisms many decades later. The work of Reeves and Nass (1996) highlights that there may be an avenue of research available to researchers for implementing HHC concepts into technology. Interpersonal communication theories such as common ground (CG)

DOI: 10.1201/9781003213963-7

(Clark, 1996) and the cooperative principle (Grice, 1975) have given researchers ways of approaching interface design considering how operators give and receive information during coordinated processes (e.g., Eriksson & Stanton, 2017c).

Clark and Brennan (1991) outline the concept of 'CG' in which language is a collaborative process where CG is established through 'joint activity' using preexisting CG. Clark's (1996) hypothesis describes how humans utilize CG as a way of 'minimizing effort' during collaboration and that interactions should be made explicit to ensure that both parties are coordinating sufficiently. Even though this theory originates from HHC research, the concept has been applied to CMC and HCI. CG implicates methods of communication that involve bidirectional interaction and feedback, as well as visual references that better establish CG, to be useful in communicative processes.

Similarly, communication is more effective if it aligns with the four maxims outlined by Grice's cooperative principles (1975):

1. The Maxim of Quantity—The degree to which information is informative but not overloading.
2. The Maxim of Quality—The degree at which information is grounded in truth and how well it is supported by evidence.
3. The Maxim of Relation—The degree at which information is relevant to the task/activity being conducted.
4. The Maxim of Manner—How well information is provided in relation to ambiguity, obscurity, and presentation, that is, briefly and orderly.

These maxims have been applied to automation interaction in cockpits and in automotive design (Eriksson et al., 2017; Eriksson & Stanton, 2018) and have been proposed as a useful tool in automation design. As Grice's cooperative principles (1975) provide specific guidance for information transfer, these maxims will be used to evaluate content within each communication strategy. The principles serve as a useful tool for initial stages of interaction design, as content and method of communication can be addressed before future chapters begin to refine these initial interactions with additional theoretical approaches. However, the maxims are not specific in their recommendations and do not provide the depth required to fully inform this book's outcomes.

Many of these factors have since been applied to vehicle automation (e.g., Bickmore & Cassell, 2001; Eriksson & Stanton, 2017c). Notably, Klein et al. (2004, 2005) develop Clark's (1996) concepts to be more applicable to modern automated systems showing that HHC can be effectively applied to human–machine teams. They show how joint activity is the product of a range of factors including CG, interpredictability, communicating capacity information, and signaling states and phases. Further, to apply the concepts of interpersonal communication to technological systems, 'natural voice recognition' has been proposed as an important component in future driver–vehicle coordination (Harvey & Stanton, 2013; Large et al., 2017). The use of vocal interaction during handover is typical in a range of high-risk domains such as medicine, aviation, and energy manufacturing/distribution (Patterson et al., 2004) and has been proposed as a method of information transfer in C/HAVs

(Bazilinskyy & de Winter, 2015; Eriksson, & Stanton, 2017c). This interaction style is supported by the literature surrounding capacity limits in working memory, which shows that multiple channels exist to process different modalities of information (Baddeley, 1992). Research into the use of multi-media suggests that the presentation of vocal cues better compliments visual information than the use of pictorial cues, as pictorial cues may contend with other visual cues for attentional resources. This suggests that there may be advantages to using audio and vocal communication when compared to visual interfaces alone (Bazilinskyy et al., 2015; Gyselinck et al., 2008; Large et al., 2017).

In C/HAV handover, knowledge of what information should be relayed to the driver and how this should be done is limited. The handover task has been examined in shift-work domains such as healthcare, aviation/air traffic control, and energy manufacturing/distribution. In these domains, failure to transmit information effectively during handover can lead to disastrous consequences (Patterson et al., 2004; Salmon et al., 2009). Strategies of communication in these domains could inform C/HAV handover, as they both involve a human taking control and responsibility for a task from another agent, including what information is required, how information is prioritized, and how it is delivered (e.g., Adamson et al., 1999; Eurocontrol, 2012; Patterson et al., 2004; Riesenberg et al., 2009a). We therefore propose that these preexisting strategies in these domains should serve as a basis for testing handover protocol in C/HAVs.

Handover in these domains is typically conducted between two human shift-workers due to changing personnel. Vocal communication is the primary method of information transfer in many domains (Patterson et al., 2004). Most notable strategies across the literature are the requirement for a structured checklist (e.g., Riesenberg et al., 2009a), vocal bidirectional exchange (e.g., Rayo et al., 2014), good knowledge of the past (e.g., Adamson et al., 1999), training (e.g., Li et al., 2012), and technological aids (Cheung et al., 2010). These strategies allow the person taking over to establish their own mental model of the driving environment (Revell & Stanton, 2012; Stanton & Young, 2000). All of these methods provide a basis upon which to design handover in C/HAVs.

5.1.2 Current Study, Aims, and Hypotheses

The aim of this study was to create a handover environment analogous to that of shift handover. In this setting, two drivers can vocally communicate with one another and exchange control of the vehicle. In doing so, preferred handover strategies and the types of information that were transmitted during handovers were measured. Grice's (1975) cooperative principles are applied to the experimental handover structures to guide discussion. The aims of this research were:

- To assess workload, usability, and acceptance concerning tested handover methods.
- To gain a better understanding of drivers' naturalistic (Angrosino, 2016) and preferred information content and method of information transmission during handover.

- To assess workload, usability, and acceptance concerning tested handover methods.
- To gain a clearer understanding of why drivers prefer or require certain types of information transfer or interaction style.
- To explore whether handover methods and information transfer has an effect on driving performance following the handover.

To achieve these aims, the following hypotheses were generated:

1. When able to naturally handover to one another, with no preset structure, information transmission will increase after experiencing a set of pre-defined handovers.
2. When able to naturally handover to another driver with no preset structure, information content and methods will more likely represent that of the pre-defined conditions after taking part in them.
3. There will be a difference in driver workload (NASA Task Load Index—NASA-TLX; Hart & Staveland, 1988), usability (System Usability Scale—SUS; Brooke, 1996), and acceptance ratings (System Acceptance Scale—SAS, van der Laan et al., 1997) in relation to predefined conditions undertaken.
4. Drivers' longitudinal and lateral control will vary following handover in relation to the handover condition undertaken.
5. There will be individual differences in driver preferences to handover interaction.

5.2   METHOD

5.2.1   Participants

The study was granted ethical approval via the University of Southampton's ethics committee (ERGO No. 26691). Participants were recruited through advertisements on the university website and advertisements on campus. Forty participants were recruited aged 18–61 (29 M, 11 F; mean age = 31.1, SD = 10.07) and took part in the study. Participants held full driving licenses and had no impairments preventing the operation of the driving simulator. As the end user will be varied in driving experience, no specific driving experience criteria was set. Participants had a mean of 7,169 annual miles, ranged between 0 and 20,000 (SD = 5151 miles). As the experiment required two participants, they were paired according to availability and on a first-come-first-serve basis. Adverts asked participants if they were willing to be paired with strangers, and this was ensured.

5.2.2   Experimental Conditions

In line with literature surrounding handover in high-risk domains (Riesenberg et al., 2009a), a structured checklist was constructed inspired by two concepts: IPSGA (information, position, speed, gear, acceleration) as a driver coaching system (Stanton et al., 2007) and PRAWNS (pressure, roles, airports, weather, nonstandard information, strips to display; Walker et al., 2010; Wilkinson & Lardner, 2013). The design of the checklist featured a half-day discussion with seven human factors experts (5M, 2F) all working within the C/HAV domain. These experts were tasked with considering the major features of returning to the loop and constructing a checklist like that found within domains. The workshop worked on a divergence and convergence approach, by suggesting a wide variety of factors and then converging on six items for inclusion (closely replicating the item count within IPSGA and PRAWNS). The resulting checklist served as a starting point for testing and future work. The checklist proposed was called 'HazLanFueSEA' and represents:

- Controls—Instruction to place hands/feet on wheel
- Hazards—Information such as close vehicles
- Lane—Lateral position on the roadway
- Fuel—Indicated as 'miles remaining'
- Speed—Indicated as 'miles per hour'
- Exit—Information on junction number and distance from junction
- Action—The action the driver is required to take (e.g., enter left lane and exit)

HazLanFueSEA starts with a request for the current driver to place their hands on the wheel and their foot on the accelerator prior to receiving the checklist information.

To explore driver behavior, attitudes, and the effects on natural handovers, four 'predefined' handovers were tested, inspired by the literature in shift handover. These conditions were implemented using a repeated measures design:

- Outgoing driver delivering HazLanFueSEA with read-back responses from the incoming driver (see Boyd et al., 2014). Referred to in this study as 'Checklist with Read-Back' (CL/RB).
- Delivering HazLanFueSEA in an interactive conversation-style questioning format conducted by the current driver (e.g., Question: what is in the left lane? Answer: an approaching red car; Bickmore & Cassell, 2001). Referred to in this study as 'Checklist with Guided Questioning' (CL/GQ).
- The questioning of the current drivers' knowledge and intentions, asking whatever the incoming driver feels is necessary. This condition was inspired by CG theory and the presence of a two-way process that can foster the repair process in communication. Referred to in this study as 'Open Questioning' (OQ) (see Clark & Brennan, 1991; Rayo et al., 2014).
- A timed handover involving no information transfer regarding the driving environment (60 s countdown) modeled on an existing handover design revealed by Volvo (Volvo Car Group, 2015). Referred to in this study as 'Timed' (T).

Before delivering CL/RB, CL/GQ, and OQ, the driver in control of the vehicle initiated the handover process by stating to the other driver 'I am ready to handover'. The driver in control awaited an acknowledgment prior to delivering the vocal protocol specified in the condition.

To guide our discussion, conditions were rated in relation to Grice's (1975) maxims of effective communication (see Table 5.1). Cohen's kappa was run to determine whether there was agreement between two researchers' ratings regarding how well the conditions address Grice's cooperative principles. There was moderate agreement between the two researchers' ratings ($k = .464$, $p = .019$).

- Quantity—The checklist conditions provide a lot of information but still run the risk of over-providing information, whereas open questions are dependent on whether the driver asks for information. But to them, they will get a subjective optimum quantity of information.
- Quality—CL/GQ encourages the search for a driver's own evidence. CL/RB and OQ may be in part dependent on faith, but information is still in the environment/on the cluster.
- Relation—Checklist items are assumed to be relevant to the task, but this is yet to be confirmed, whereas open questions meet this maxim by only providing what the driver desires.
- Manner—The CL/RB condition is the most structured and orderly, followed by CL/GQ. Open questions may be less prescriptive and result in confusion as to the original question asked.

It is worth noting that the timed condition does not deliver any explicit scenario information, so ratings for timer are not provided for the first three maxims. However, as the structure is orderly, it scores highly on the maxim of manner.

5.2.3 Design

These predefined conditions were counterbalanced across participants using a balanced Latin square where conditions occurred at different times and were preceded and followed by a different condition in each combination (see Figure 5.1). Ten participants took part in each combination.

To test whether naturalistic information content and transmission methods were influenced by the participation in predefined handovers, two 'free-form' handover conditions were conducted before and after taking part in the four predefined

TABLE 5.1

Authors' Ratings of Predefined Vocal Structures in Relation to Grice's (1975) Maxims

	CL/RB	CL/GQ	OQ	Timed
Quantity	3(?)	3(?)	4	—
Quality	3	5	3	—
Relation	3	3	5	—
Manner	5	4	3	5

Note: Ratings range from 1 to 5. (?) Indicates an uncertainty, particularly whether there is too much/too little information being delivered due to individual preferences

1	A	B	C	D
2	B	D	A	C
3	D	C	B	A
4	C	A	D	B

FIGURE 5.1 A balanced Latin square showing the counterbalancing of predefined conditions.

conditions, following an ABA design (where A denotes a free-form handover condition and B denotes the four predefined handover interactions). In the 'free-form' conditions, participants could engage in the handover however they wished and were not given any indication as to what information to transmit or how to transmit it, so that emergent themes from 'free-form' handovers can be compared to predefined handovers experienced and gauge whether they had an influence on 'free-form' handover interactions. After the experiment, participants filled out a short questionnaire asking them to provide their thoughts on the predefined conditions and give any of their own recommendations for future handover design.

The independent variable was handover condition. The dependent variables were lateral and longitudinal inputs, information vocally transmitted and method of transmission (applicable only to unstructured conditions and open questions), subjective workload, acceptance, and usability.

5.2.4 APPARATUS

The study was conducted in the Southampton University Driving Simulator (SUDS). The simulator has a fixed base and 135-degree field of view. The simulator is designed to create a safe environment from which to analyze driving behavior that does not incur the adverse effects of distraction on the road yet still aims to create a naturalistic environment. Validity comparisons between simulator and real-world environments show that a simulated environment can produce strong positive correlations in driver behavior (e.g., Eriksson et al., 2017). However, limitations such as reduced risk perception have been demonstrated in such environments (Underwood et al., 2011). The simulator was modified so that the Land Rover Discovery could support two steering wheels, two displays, and two sets of control inputs. The steering wheels had the capability of taking control of the vehicle through the click of a button on the device taking control (see Figure 5.2).

STISIM Drive was used to create a freeway environment modeled on the M24, junction 5–7, in the UK. Assuming an average speed of 60 mph, the constructed route environment was looped two and a half times during a single scenario. This was to ensure that the condition ran for enough time for six complete handovers to take place with enough time for the driving environment to change prior to the next handover (10–12 min). To balance traffic density, traffic speed, and headway, traffic was set at 52/57/62 miles per hour in the left, middle, and right lane, respectively. This allowed cars to be placed closer to one another without their reactive behavior

FIGURE 5.2   Driving simulator setup to simulate handover between two drivers.

being triggered (i.e., when there is less than a 3 s headway to an obstacle), which would result in them braking and causing congestion. Cars were then placed within a minimum of 300 feet in front of one another to clear this headway time. They were then varied randomly up to 600 feet. Traffic was generated after 1,000 feet to ensure the participant could match their speed before its appearance.

5.2.5   PROCEDURE

After reading the information sheet outlining risks such as motion sickness, the participant provided their informed consent. Following this, participants were introduced to the procedure and instructed on how to operate the driving simulator. This included an introduction to the visual display and the button they were required to press when taking control. A trial was then conducted consisting of 5 min of motorway driving (two and a half each) to allow participants to get familiar with controls and the environment. No vocal handovers were conducted, and control was switched when the experimenter instructed.

Participants were then shown how to fill out the three questionnaires following each condition, the NASA-TLX, the SUS, and the SAS. When participants were ready, experimental blocks began. The study design was a repeated measures design, with each pair taking part in each handover procedure. There were six handover conditions. The first and last conditions were always natural, where participants were allowed to handover to one another as they wished, with the four predefined conditions being conducted in the specific counterbalance order for that pairing.

Participants were given an example script of how the condition would take place and a cue card was attached to the steering wheel to remind them of the current condition (see Appendix A for cue cards for each condition). Participants were asked to simulate a motorway junction exit of junction 7, where the action was to move into the left lane exit. Each condition consisted of six handovers (three in either the role of automation or driver) and took approximately 10 min to complete. Each handover consisted of participants transferring control to one another, initiated by the person currently in control of the vehicle. Each moment a handover was to be performed was dictated by the experimenter tapping the participant's shoulder. Handovers were spaced approximately 1 min apart although variance at the experimenter's discretion was conducted to avoid prediction. Thirty-six handovers were conducted throughout the entire experiment. The entire procedure took between 1.5 and 2 h to complete depending on speed of questionnaire completion. Participants were given short breaks between conditions whenever they felt they needed them.

Once the experimental blocks had been complete, participants filled out a questionnaire where they were able to provide feedback on the experimental conditions and provide their suggestions for future handover design. They were then debriefed, paid £10 to cover travel expenses, and thanked for their time.

5.2.6 METHOD OF ANALYSIS

Transcripts were coded according to the information vocally transmitted and method of delivery during handover. For each participant the percentage of handovers (out of six) involving the transmission of each information type and each method of transmission was calculated for each pair. A t-test was conducted to analyze whether information transfer increased from before to after taking part in the predefined handover conditions. The same percentage calculation was used for the 'open questioning' condition to measure what information was transmitted during this condition.

As questionnaire results were generated from a repeated measures design, the analyses undertaken were Friedman tests with Bonferroni-corrected Wilcoxon post-hoc analyses (six comparisons; $\alpha = .0083$). The longitudinal velocity data represented a parabola pattern; therefore, this data was analyzed using a linear regression with a quadratic term to test intercept, slope, and shape. Finally, to give a good indication of the extent at which drivers veered and corrected themselves following handover, lateral velocity data were averaged across 5 s following the handover and analyzed using a Friedman test followed with Wilcoxon post-hoc analyses (six comparisons; $a = .0083$).

5.3 RESULTS

5.3.1 'FREE-FORM' CONDITIONS

Figure 5.3 shows the mean percentage of handovers for all participants that involved information transfer before and after experiencing the four predefined conditions. The percentage sharply increases, indicating a higher likelihood of information transfer during the second 'free-form' condition. A t-test indicated a significant difference ($t(19) = 5.8, p < .001$).

Information transmitted during handover greatly increased across all categories except for information types: 'adapt action', 'advice giving', and 'road layout', which decreased in usage slightly during the second 'free-form' condition (see Figure 5.4). The most common information themes to appear in the first 'free-form' condition were speed, other vehicles, and vehicle position. There were also occurrences of instructing one another to change speed/position (adapt action) prior to handing over. Information types with no bar present either before or after predefined conditions representing zero instances of the transmission of this information type over all experiments.

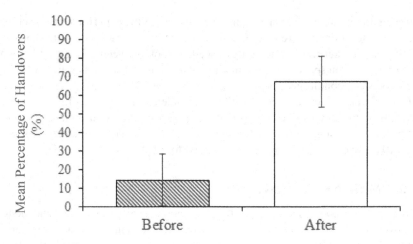

FIGURE 5.3 Mean percentage of handovers involving information transmission, before and after predefined conditions.

In the second 'free-form' condition, the most common information types were those transmitted during the predefined condition. There were many new information types transmitted that were unrelated to the predefined conditions such as weather/road conditions, traffic information, instruction giving, and asking about/challenging intentions. Drivers transmitted a mean of 0.39 (SD = 0.95) information types before predefined conditions, increasing to 1.6 (SD = 1.67) after the predefined conditions. These information types were grouped based on prevalence and represent four distinct groupings (Figure 5.4).

There was a great shift in methods used before and after the predefined conditions, these methods are defined in Table 5.2.

Before the predefined conditions, the majority of handovers consisted of a simple exchange, representing only a handover notification and a confirmation with no information transmission. Following the predefined conditions, participants adapted their strategies through either combining methods or staying with a single handover method. Figure 5.5 shows the mean percentage of handovers using a particular handover method, both before and after predefined conditions.

Figure 5.5 shows the percentage of handovers that involved method hybridization, increasing from 4% to 25%. Simple exchange decreased from 85% to 30%. All other methods, with the exclusion of read-back, which did not get used in either 'free-form' conditions (represented by the absence of bars in the bar chart) increased in usage. The most common method was open questions, although this method was commonly hybridized with other methods. One method, that was not used in the predefined conditions is categorized as 'monologue' and represents the one-way delivery of information in a single packet of speech. Further, when CL/GQ was utilized it typically only included information regarding hands/feet on control inputs. Simple exchange remained one of the most common methods of handover and reflects the percentage of handovers that did not involve information transfer (see Figure 5.5).

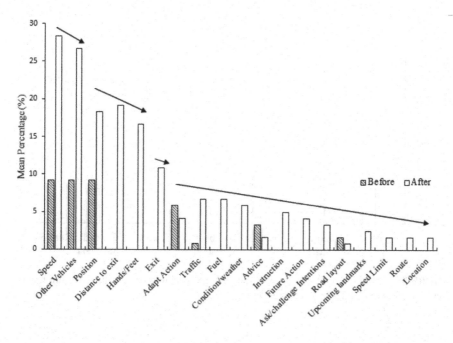

FIGURE 5.4 Natural conditions—mean percentage of handovers including information transmission for each information type before and after predefined conditions.

TABLE 5.2
Definitions of Handover Methods Conducted during 'Free-Form' Handover Conditions

Handover Method	Definition
Hybrid	The use of two or more of the following handover methods combined in a single handover
Simple exchange	Handovers that consisted only of a notification and a confirmation, with no information transmission
Monologue	Handovers that consisted of a steam of information being delivered from the current driver, with no response being asked for from the incoming driver, such as that of CL/RB or CL/GQ
Open questioning	Handovers that involved an interaction on the basis of questioning the current driver about what they know about the driving environment
Timed	The use of a countdown conducted by the current driver to indicate a timeframe for receiving control. Deployed with or without information transfer
Guided questions	The use of interactive questioning, where the current driver quizzes the incoming driver regarding the driving environment
Read-back	Handovers that use a repeat-back method to demonstrate that information is being received

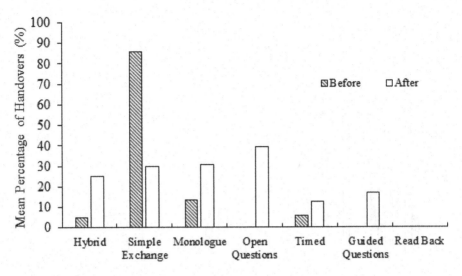

FIGURE 5.5 Natural conditions—mean percentage of handovers conducted using handover method before and after predefined conditions.

5.3.2 USE OF OPEN QUESTIONING

During the OQ condition, drivers seemed to interact in a different manner. Questions revolved around immediate threats, speed, and future action, with a specific focus on what will come up ahead, such as traffic, route information, and whether there any upcoming landmarks such as petrol stations or speed cameras. Participants transmitted a mean of 2.05 (SD = 1.27) information types per handover interaction. Information types transmitted during this condition are displayed in Figure 5.6.

5.3.3 NASA-TLX, SUS, AND SAS

The assumption of normality was violated for multiple combinations of condition/dependent variable, therefore nonparametric tests were conducted. Four values were missing due to participant error. As there were so few missing values it seemed unnecessary to remove cases in full, therefore these values were imputed using the expectation maximization method (Borman, 2004). Descriptive statistics are displayed in Figure 5.7.

Four Friedman tests were conducted for each dependent variable: NASA Task Load Index scores, System Usability Scale scores, and scores from the two subscales of the System Acceptance Scale. These were analyzed using handover condition as the within-subject variable with four levels: read-back, CL/GQ, open questions, and timed. A main effect of handover condition was found for all dependent variables tested ($p < .05$). Results are displayed in Table 5.3.

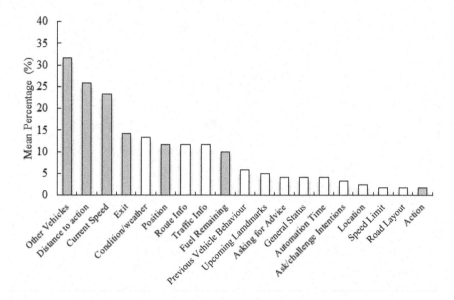

FIGURE 5.6 Mean percentage of handovers including information transmission for each information type used within the 'open questions' condition. Grayed bars indicate information transmitted in HazLanFueSEA.

Table 5.4 displays post-hoc pairwise comparisons for questionnaire responses. Significant differences vary based on measure, although typically, these comparisons can largely be grouped into read-back and CL/GQ receiving similar mean ratings, as well as open questions and timed. Differences were greater in the NASA-TLX and SUS, whereas the AS Usefulness scale showed few significant differences during post-hoc analyses.

5.3.4 Change in Speed following Handover

Figure 5.8 displays mean speeds following handover over the first 5 s of taking control. All conditions showed a parabola effect where speed decreased in the first 2 s following handover with speed steadily increasing from between 2 and 4 s following handover.

Table 5.5 shows the output for a linear regression using the stepwise enter method. As the data appear to be nonlinear, speed as a function of time did not show an effect. However, adding the condition intercept and the quadratic terms both improved the model ($p < .001$) showing that conditions significantly differ in their overall speeds. The model of best fit as a result of linear regression stepwise analysis was 'Speed ~ Time + Cond Intercept + Quadratic Term'.

5.3.5 Lateral Velocity following Handover

Mean lateral velocity was generated by taking the first 5 s following handover, calculating the square root of squares, and averaging across the time period, illustrated

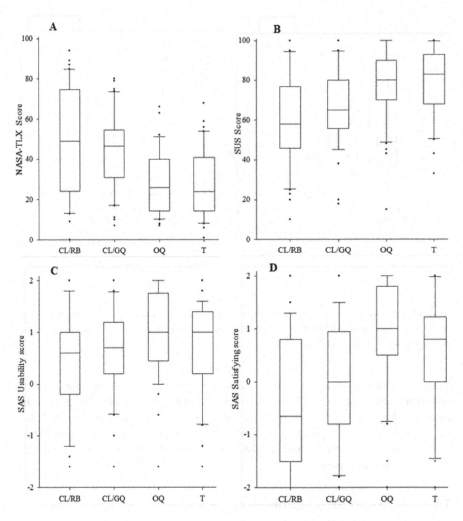

FIGURE 5.7 Box plots to show results from NASA-Task Load Index (panel A), System Usability Scale (panel B), System Acceptance Scale—Usability (panel C), and System Acceptance Scale—Satisfying (panel D) between the four predefined handover condition.

TABLE 5.3
Results from the Four Friedman Tests Analyzing Mean Differences in NASATLX, SUS, and the AS Subscales between Handover Conditions

Source	df	χ^2	p
NASA-TLX	3	34.64	.001*
SUS	3	26.01	.001*
AS usability	3	11.37	.01*
AS satisfying	3	24.14	.001*

*p < .05

TABLE 5.4
Post-hoc Analyses—Wilcoxon Signed Rank Tests Analyzing Differences in Scores for the NASA-TLX, SUS, and the AS Subscales between Each Handover Condition

Pairing	NASA-TLX			SUS			SAS-Use			SAS-Satis		
	Z	Sig.	r	Z	Sig.	r	Z	Sig.	r	Z	Sig.	r
CL/RB–CL/GQ	−1.08	.28	−0.17	−1.97	.049	−0.31	−1.78	.074	−0.28	−2.21	.027	−0.35
CL/RB–OQ	−3.94	.001*	−0.62	−3.93	.001*	−0.62	−2.97	.003*	−0.47	−4.08	.001*	−0.65
CL/RB–T	−4.03	.001*	−0.64	−4.46	.001*	−0.71	−1.12	.263	−0.18	−3.23	.001*	−0.51
CL/GQ–OQ	−4.41	.001*	−0.70	−2.87	.004*	−0.45	−2.01	.044	−0.32	−3.30	.001*	−0.52
CL/GQ–T	−4.20	.001*	−0.66	−3.28	.001*	−0.52	−.13	.900	−0.02	−2.04	.041	−0.32
OQ–T	−.24	.807	−0.04	−.11	.912	−0.02	−2.28	.023	−0.36	−1.68	.093	−0.27

Note: $\alpha = .0083$, * indicates significance ($p < .0083$)

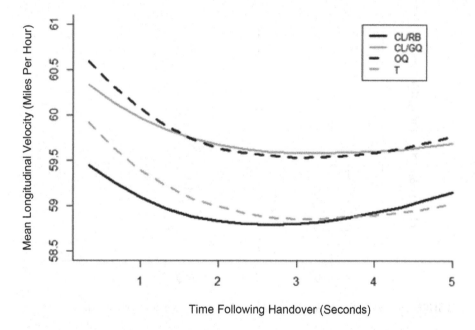

FIGURE 5.8 Line graph showing the change in vehicle longitudinal speed post-handover.

by the following equation: Eqn001.eps. Box plots for lateral velocity are shown in Figure 5.9.

A Friedman test was run to explore whether handover condition had an effect on mean lateral velocity 5 s after handing over control. The test indicated that there was a main effect of handover condition on mean lateral velocity [$\chi^2(3) = 8.2$, $p = .042$]. Post-hoc Wilcoxon signed rank tests with Bonferroni corrections ($\alpha = .0083$) revealed a significant difference between CL/GQ paired with CL/RB ($r = −0.59$, $Z = −3.24$, $p = .001$) and OQ ($r = −0.49$, $Z = −2.71$, $p = .001$)]. All other comparisons did not

TABLE 5.5
Table to Show Models Fitted to Changes in Speed Following Handover and their Associated Inferential Statistics

Model	df	F	p	SST	Δr^2
Speed ~ time	1, 2,278	—	.077	—	.0009
Speed ~ time + cond intercept	1, 2,277	65.87	.001*	9	.0006
Speed ~ time + cond intercept + quadratic term	1, 2,276	748,004	.001*	98,070	.997
Speed ~ time * cond intercept + quadratic term	1, 2,275	1.66	.2	0	.997
Speed ~ time * cond intercept + quadratic term * cond	1, 2,274	0.25	.62	0	.997

Note: Cond = Condition, * indicates $p < .05$

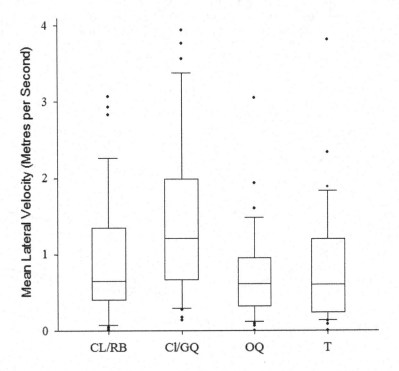

FIGURE 5.9   Mean lateral velocity over 5 s following the handover of control.

show significant differences ($p > .0083$). An illustration of how this relates to time is displayed in Figure 5.10.

5.3.6   QUALITATIVE FEEDBACK

To explore why drivers preferred certain strategies to others, qualitative feedback was received regarding the predefined conditions. Positive and negative feedback related to the predefined conditions are summarized by the themes outlined in Table 5.6.

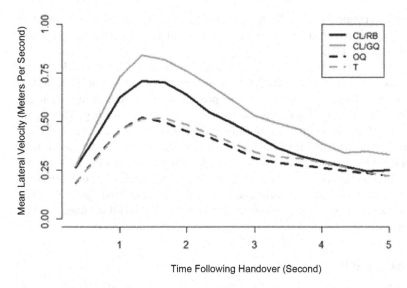

FIGURE 5.10 Line graph showing lateral velocity over 5 s following the handover of control split across handover conditions.

TABLE 5.6
Themes Generated from Qualitative Questionnaire Responses with Regards to the Four Predefined Conditions

	Themes	CL/RB	CL/GQ	OQ	T
Positive themes	Safety	9	8	6	—
	Efficiency	3	2	3	6
	Personalization	—	—	10	—
	Self-paced	—1	—	—	3
Negative themes	Inefficient	8	8	3	—
	Frustrating	9	5	3	—
	Unsafe	—	—	—	8
	Stressful	—	—	—	7
	Unnecessary	6	—	—	—
	High workload	—	—	2	—

Participants generally acknowledged that CL/RB, CL/GQ, and OQ played a role in safety (e.g., Participant 6—'Repeating back is useful, as you become aware of the situation'). Some participants felt that the timed condition allowed for efficiency (e.g., Participant 39—'The timed condition was simple and quick'), although some felt the same about the other conditions. The highest number of responses in a given condition was the personalization of OQ, allowing contextual information to be relayed that the driver thought to be most useful (e.g., Participant 18—'Being able to question was good as I found out what I needed to know'). However, a few found benefits in being able to conduct the handover alone using the timed condition without being dictated to (e.g., Participant 15—'Timed made me feel confident that I could assess the situation for myself').

Negative feedback reflected themes measured in the questionnaires. Most comments related to CL/RB and CL/GQ were concerned with the verification method rather than the information the system was providing. For example, many participants found CL/RB to be frustrating, inefficient, and unnecessary (e.g., Participant 7—'Repeating back was annoying and time consuming'); however, many of these comments were also made for CL/GQ and OQ (e.g., Participant 4—'Questions took too much mental toll, took a while to learn and get used to, and were most annoying'). The timed condition received no feedback regarding inefficiency or frustration, instead participants found it stressful and largely unsafe (e.g., Participant 19—'The timed one was easy, but I felt pressure to take over quickly so I don't always check all the info').

5.4 DISCUSSION

The aims of this research were to explore how vocal communication during handover can be applied to C/HAVs. Two drivers handed control between one another in a driving simulator equipped with two steering wheels and two sets of control inputs. In particular, this study explored naturalistic and emergent handover vocal communication methods (Angrosino, 2016; Eriksson & Stanton, 2017c) and the information types transmitted during 'free-form' conditions. This study found that information transmission increased and vocal communication methods and information transmitted differed greatly from before to after experiencing a set of predefined conditions inspired by shift-handover literature. Predefined conditions consisted of a CL/RB (Boyd et al., 2014), a checklist with conversation-style questioning (adapted from Bickmore & Cassell, 2001), OQ of current driver (adapted from Rayo et al., 2014), and a timed handover with a countdown. Differences in vehicle control between predefined conditions were also found, and qualitative feedback provided insights into drivers' preferred handover protocols.

Overall, the percentage of 'free-form' handovers involving information transfer increased dramatically after experiencing the predefined conditions. The reasons for this change may include drawing inspiration from the predefined conditions, as well as learning effects. Regardless of the contributors to this change, this highlights areas for consideration, such as the use of training programs, which reflects a current method undertaken in shift handover (e.g., Li et al., 2012).

Due to interactions featuring primarily vocal information, Grice's (1975) maxims could provide valuable insight in how to interpret these findings. In this section, each maxim is discussed in relation to the interactions that took place and the feedback provided by participants regarding each condition.

- Quantity—In the CL/RB and CL/GQ condition, participants found that the communication method was inefficient, thereby violating the maxim of quantity. It appears that participants did not feel as if it was necessary to verify information transfer for each individual piece of information transmitted (not overloading; concise delivery of information). For the OQ condition, participants felt that the level of personalization was a positive aspect, indicating that they prefer to dictate the amount of information provided themselves.

- Quality—It was thought that the requirement to check information visually in the CL/GQ condition would be beneficial to drivers. However, our findings do not provide any insight into whether drivers felt that the truthfulness of either condition was greater than another.

- Relation—In this study, the maxim of relation and quantity appear to be closely linked. Information deemed to exceed the optimum quantity could also be assumed to be unnecessary for the task. The procedure involved in the CL/RB conditions was deemed as unnecessary for several drivers, as well as scoring low on the acceptance/usability scales, whereas the comments regarding the high degree of personalization offered by OQ and the ability to ask questions that are contextual indicates that OQ addresses this maxim sufficiently. OQ of the current driver was the most common method observed, although this was likely to occur in conjunction with other methods. This shows that, when available, drivers naturally interact with the automated system to provide them information that they feel is necessary. These findings support the usage of a two-way interaction system, much like that of shift handover in other domains (e.g., Cheung et al., 2010; Patterson & Woods, 2001).

- Manner—Subjective questionnaire ratings show that CL/RB and CL/GQ conditions may not be the preferred way of communicating with an automated system. Rather, participants would more likely establish whether transmission was successful following the complete delivery of information (the strategy of monologue; Figure 5.5, Section 3.1) or favor nonverbal exchanges (i.e., simple exchange; Figure 5.5, Section 3.1). The timed condition may indeed meet the maxim of manner, as this condition may be viewed as simplistic and concise.

Differences in the information types transmitted in the two 'free-form' conditions could also provide insights into what information participants would like to receive from vocal communication with automation. Before taking part in the predefined conditions, the most common information types were speed, other road vehicles, position on road, as well as requests being made between both participants to adjust their driving behavior prior to handover. Initially, these findings confirm the intuitive nature of our checklist, as speed, other vehicles, and lane position are featured in HazLanFueSEA and the inspirations for its development (Walker et al., 2010; Stanton et al., 2007). These information types remained in the top four most transmitted types after the predefined conditions were conducted. Drivers' willingness to interact in this format provides further evidence toward the effectiveness of using checklists in the handover task, at least as a training tool (Riesenberg et al., 2009a; Walker et al., 2010).

All but one information type increased after taking part in the predefined conditions, showing an increase in the diversity of information transmitted. Interestingly, information typically revolved around HazLanFueSEA, with additional information types including traffic/route information and weather conditions. These novel information types should be considered for future checklist design (Riesenberg et al., 2009a; Walker et al., 2010). The least common information type from HazLanFue-SEA was the action element. One interpretation of this is that moving to the left lane is implicit knowledge, and so in our scenario this would not be an information type necessarily worth transmitting but may be more relevant in other contexts.

Findings related to longitudinal behavior post-handover show that there is no difference in velocity change as a result of handover method. Conversely, mean lateral velocity over time, following handover, seemed to be a lot higher in the CL/GQ condition. Explanations for this include the requirement for the driver to pay attention to the environment to answer questions, whereas the CL/RB, OQ, and T conditions did not require the driver to actively search the driving environment. The effects of handover interactions and visual behavior could be a consideration during future experiments.

From a design perspective, a balance should be struck between the system being usable and the system being safe. Participants reported that the HazLanFueSEA conditions were safe due to the wealth of information being transmitted. However, drivers generally prefer OQ due to its flexibility in providing information the driver wants to know as well as allowing the system to acknowledge that the user is engaged. This supports a handover design inspired by two-way communication (e.g., Cheung et al., 2010; Patterson & Woods, 2001). This may have its own setbacks, as a potential issue for designers could be that OQ does not deliver need-to-know information unless asked for. Further, as 30% of handovers were that of simple exchange to increase safety, dynamic interfaces could be used to compensate for the reduced interactions that take place in this condition. In doing so, drivers that prefer not to engage in vocal communication can receive handover assistance through another medium (e.g., Tonnis et al., 2005; Walch et al., 2015). Finally, attention should be paid toward how the system confirms the driver's SA. That is, if legal or safety concerns invoke attention in this form of handover design (de Carvalho et al., 2012).

One of the key findings of this study is the diverse nature of drivers' preferences. During experiments, drivers either preferred to interact vocally in great depth or handover with limited amount of information transfer. Usage of information types was diverse with many information types being utilized by only a few drivers. To that end, the results from this study point toward the need for personalization and integrate the features that drivers find to be most useful across all conditions (see Small et al., 2011; Weld et al., 2003).

There have been a number of suggestions for human–machine interface (HMIs) to provide a platform for information relay during handover, as well as providing information to the driver during automated driving, representing the approach of combined performance of driver and automation (Merat & Lee, 2012). To our knowledge, how vocal interaction with can be integrated with visual cues regarding handover is yet to be considered in detail. Many of these streams have been considered in isolation, for example, handover assistants (Walch et al., 2015), vocal feedback (Eriksson & Stanton, 2017c; Stanton & Edworthy, 1999), and haptic feedback (Petermeijer et al., 2017a, 2017b). In line with malleable attentional resource theory (MART) the attentional resources of a human is finite and variable across certain tasks (Wickens, 1991; Young & Stanton, 2002a, 2002b, 2004), therefore designs should be tested that combine already tested concepts in handover design so that a single effective handover protocol can be formulated and tested.

5.5 CONCLUSIONS

This study explored the use of vocal communication as a tool for handover HMI design in level 3 and 4 vehicles (Large et al., 2017; Stanton & Edworthy, 1999). How this technology can be applied is currently not known. This study addressed the application of vocal communication in the handover task in the context of level 3 and 4 vehicles by conducting handovers in a range of 'free-form' and predefined conditions in a dual-controlled driving simulation between two human drivers.

Naturalistic and emergent themes from driver communication indicate an openness to information transfer, including the use of a checklist and interactive questioning. Due to the changing handover behavior from before to after experiencing structured conditions, these results provide evidence for the potential effectiveness of training programs to encourage effective vocal handover styles. Being able to ask questions of the automation was also the most utilized in free-form conditions and was rated the most usable, accepted, and least mentally demanding process. If constraints allow for such designs, this approach may be fruitful in future design.

Qualitative feedback and questionnaire data indicate the requirement for further examination of SA verification methods, as well as the requirement to explore the potential for personalization suited to driver preferences. Results also indicate that HMI designs should have the capacity to provide contextual information that is tailored to the environment. However, if a system must ensure information transfer has occurred, more exploration into driver-to-vehicle feedback is required as participants in this study demonstrated frustration to conditions such as read-back. Overall, drivers desire an efficient, safe, and usable handover assistant. We propose that further research should determine whether the methods and information types generated by this experiment do indeed raise SA prior to handover and how personalization can be applied to handover HMI design.

As a result of this experiment, the following should be considered in future vocal handover assistant designs:

- A usable and efficient way to confirm information transfer so that drivers do not become frustrated with handover interactions.
- The delivery of crucial and concise information, so that drivers receive the information they require without unnecessary information being received.
- A degree of personalization to facilitate individual differences/preferences.
- A way for drivers to gain up-to-date contextual information on demand.

5.5.1 FUTURE DIRECTIONS

This chapter proposes that communication within human teams can be learned from in order to design communicative handover assistants in conditionally and highly automated vehicles (C/HAVs). An experiment tested four vocal strategies in handing over control, with variability in how situation awareness is raised. Findings suggest that questioning the automation prior to handover (analogous to that of clinical teams) may be the most usable and acceptable methods for communicating information. A large proportion of handover research suggests that multimodal cues are important in alerting and communicating with the driver, with an understanding of how and what should be vocally communicated during handover. Chapter 6 explores how visual displays may be used in conjunction with vocal guidance and explore which displays are more effective in addressing.

6 Directability and Eye-Gaze

Exploring Interactions between Vocal Cues and the Use of Visual Displays

6.1 INTRODUCTION

So far, C/HAV interaction strategies that feature flexible and contextual handover communication seem to be more effective in addressing target outcomes. Now, with a good understanding of domain values, physical properties, and potential applications to the domain, this chapter turns attention toward how visual information can supplement vocal communication for developing collaboration between driver and vehicle throughout the automation cycle. Before entering the design stage of this book, this chapter builds on Chapter 5 (Clark et al., 2019b) by analyzing data from a driving simulation featuring a handover assist prototype developed by the Human Interaction: Designing Autonomy in Vehicles (HI:DAVe) project. The data extracted and displayed in this chapter focuses on how the driver perceives visual information during the handover while engaging with vocal interaction. With a combination of visual, vocal, and physical input modalities, a handover assist will be maximized in dealing with domain values, as outlined by the cognitive work analysis chapter (Chapter 3). The outcomes of this chapter will allow for prototypes to present visual information to the most effective visual modality. For C/HAV handover, this could include directing the driver toward information such as hazards in the environment, the status of the vehicle, and upcoming events or actions. As highlighted in Chapter 2, automation assistants designed in this way may go some way to alleviate the vulnerabilities that arise during the transfer of control. These assistants provide the driver with information and guidance as to what is going on in the environment, what is expected of them, and when/how transitions will take place. Previous research also indicates that vocal interaction style may have a desired effect on driver visual gaze behavior (Clark et al., 2018).

6.1.1 VISUAL GAZE AND AUTOMATED DRIVING

This study was concerned with the ability for vocal interfaces to raise the awareness of a driver during automated driving via vocal interaction with the vehicle. Intuitively, access to visual information during the driving task is crucial to optimal driving performance (Owsley & McGwin, 2010). Eye movements have also been proposed as a way of measuring situation awareness and predicting task performance (de Winter

DOI: 10.1201/9781003213963-8

et al., 2018). It follows that driver eye movements during automated driving should be considered when assessing various human–machine interface (HMI) implementation. The study was concerned with a number of additional factors that may influence visual gaze during AV interactions. Visual behavior during the driving task appears to differ based on a handful of demographics: 1) Gender effects show that female drivers may conduct greater amounts of visual search during driving tasks than their male counterparts (Yan et al., 2016). 2) Older drivers may experience a deterioration in visual-motor coordination (Sun et al., 2016).

Another variable that may affect takeover performance is the duration of time out of the loop (TOOTL). TOOTL can result in slower reactions and changes in gaze behavior (Feldhütter et al., 2017), so including it as a variable within this study was necessary to understand how this could affect task performance. Finally, expecting a handover may allow for more efficient visual search during transitions, leading to greater attention toward the road environment (Merat et al., 2014). Therefore, for planned handover, such as the task implemented in this study, visual gaze durations toward areas of interest such as road ahead may be higher than that of unexpected transitions. Not only is the handover task itself of importance, but the time following the takeover (where the system remains vulnerable) requires attention. This study has a secondary objective of analyzing eye-gaze directly following handover to understand how visual interfaces are utilized during this period.

To understand how visual information streams should be utilized for the purpose of handover transactions, this study implemented a handover task in a highway environment to 1) analyze how well vocal cues can guide visual gaze during handover and 2) understand the factors that may affect how and to what extent drivers rely on different visual displays during the C/HAV handover process and moments after control is regained, by recording visual gaze behavior. This was achieved by asking drivers questions regarding the driving scenario in an attempt to direct their gaze toward areas of importance (e.g., where hazards are and the status of the vehicle). By asking questions, drivers were able to feedback their responses and allow for the system to make a judgment of whether they are aware of their surroundings, therefore acting as a confirmation that they have processed the information and are aware of it. Gaze behavior was tested for group differences for four variables—age, gender, TOOTL, and car type owned.

6.2 METHOD

6.2.1 PARTICIPANTS

The quality of eye-tracking data is inherently dependent on individual differences and can result in systematic and variable error. This has been well documented and is a notable concern for eye trackers that are not head-mounted (Aaltonen et al., 1998, 135; Hornof & Halverson, 2002). Participants that appeared to be poorly calibrated (more than or equal to 50% of unaffiliated area of interest categorization, i.e., not detected by eye tracker or gaze was recorded for areas such as vehicle frame due to poor calibration) were excluded from the analysis. Of the original 65 participants, 30 were included in the analysis. During experimentation, there were no instances of simulator sickness, and one dropout was incurred due to personal health concerns.

TABLE 6.1
Demographic Spread of Drivers

Age	Mainstream (e.g., Ford Fiesta)			Premium (e.g., BMW F Series)			Totals
	18–34	35–56	57–82	18–34	35–56	57–82	
Male	5	3	1	2	6	2	19
Female	2	4	3	0	1	1	11
Total	7	7	4	2	7	3	30

Note: $N = 30$; premium vehicle drivers made up 40% of the sample

Participants ($N = 30$) were recruited through a third party in collaboration with Jaguar Land Rover (Ethics Number: ERGO Number—41761.A3). Drivers were categorized by Jaguar Land Rover's marketing team as being drivers of mainstream (e.g., Ford, Toyota, and Fiat) or premium (e.g., Mercedes, BMW, and Rolls Royce) vehicles. The breakdown of demographics and vehicle type driven is displayed in Table 6.1.

Drivers spent 23.8 years (SD = 14.2) mean years driving, and the approximate annual mileage was 12,111 miles (SD = 5,688). Recruitment exclusion criteria included irregular drivers (less than once per week), less than 2 years of driving experience, conflict of interest either with their own or a close family member's occupation/activity, susceptible to motion sickness, and pregnancy.

6.2.2 Design

This study had three research questions:

- How well does vocal interaction guide visual gaze during handover?
- How are visual interfaces utilized differently for handover and manual driving on an individual basis?
- How are visual interfaces utilized differently for handover and manual driving as a function of TOOTL, gender, age, and current car ownership?

To address these questions, participants took part in four repeated measures driving 'trials', each consisting of three handbacks (driver-to-vehicle) and three handovers (vehicle-to-driver). For each participant, half of trials were defined as being a 'shorter' TOOTL consisting of 1 min of automated control (3 min in total for each short trial). The other half of trials were defined as being a 'longer' TOOTL consisting of 10 min of automated control (30 min in total for each long trial). The entire experiment took no longer than 2 h 40 min to complete. To navigate when handovers took place, the experiment recorded time elapsed, availability of automation, current HMI state, and information regarding what stage the driver is at within the handover process. The independent variables collected included:

- Whether the condition was set to short TOOTL or long TOOTL—repeated measures—two driving trials per condition.

- Demographic information from questionnaires: gender, age of driver and current vehicle model driven—between groups measure.
- The area in which visual gaze was categorized: road ahead, head-up display, instrument cluster, and central console.
- The proportion of vocal questions relating to each area (road ahead, head-up display, instrument cluster, and central console) during handover.

The dependent variables of interest was:

- Total gaze duration (how long participants spent looking at certain areas) toward four pre-allocated areas for categorizing gaze coordinates: road ahead, head-up display, instrument cluster, and central console (see Figure 6.1 for illustration). This consisted of a recording of the coordinates of gaze at a 20 Hz resolution categorized each 20th of a second into one of the four areas of interest, giving a total gaze time for each of the four areas for each participant. This was measured for two occasions, the time taken between notification and handover (varying between participants due to individual behavior) and the 60 s of manual driving following the takeover. Gaze coordinates allowed the categorization of gaze time for each participant toward the four areas of interest.

As each participant took part in four trials (two shorter periods and two longer periods of TOOTL), participants were counterbalanced to experience TOOTL in various orders (SSLL, LLSS, SLSL, and LSLS, where S = shorter and L = longer). These were balanced across age and gender to ensure that each demographic experienced a similar number of orders throughout data collection.

FIGURE 6.1 Example of vehicle in (top) and out (bottom) of automated mode.

Vocal interaction for the handover process was inspired by previous work into non-critical handover interactions (Clark et al., 2019a). The automated assistant asked the driver to answer a range of questions to ensure that awareness had been raised. This study implemented this method by asking the driver a set of questions before control was handed over to them. Drivers reported back vocally to the vehicle answering the questions asked. Correct responses were logged, and the question was repeated if incorrect before moving on to the next question. Table 6.2 displays the questions asked, and the relevant visual HMI component in which answers resided. Questions were randomly selected and defaulted to five questions; however, following each trial, participants could raise or lower the amount of questions asked to between 1 and 10.

To understand how well vocal interaction guides visual attention this study aimed to correlate the area where the question relates to real-time visual gaze.

6.2.3 Apparatus

STISIM drive software was used to simulate a typical UK motorway environment. The simulator was built similar to that of the Range Rover Evoque, equipped with a single front-view screen, with separate wing mirrors, and an augmented display for rear-view. In automated vehicle tasks, multimodal cues have been established as being beneficial for driver awareness and driver–automation interaction (Borojeni et al., 2016; Petermeijer et al., 2017a, 2017b; Politis et al., 2015). To that end, the simulator was equipped with a variety of HMI elements. Visually, the driver was presented with a digitalized instrument cluster (size approximately in line with that of the Range Rover Evoque), center console (18 in., 16:9, installed in the direct center of the vehicle), and a head-up display (see Figure 6.1 for image of simulator).

The vehicle was also equipped with vocal and audio information streams, ambient lighting to indicate driving mode (orange for manual, blue for automated; see Figure 6.1), and a vibrating seat that was initiated when a handover was expected and when control was safely transferred to the driver.

TABLE 6.2
List of Potential Questions Asked during Handover Protocol and their Associated Visual Display

Question #	Question Vocally Presented	Associated Visual Display
1	What color is the vehicle in front?	Road view
2	What type of vehicle is in front?	Road view
3	Are you on a corner?	Road view
4	Do you see a corner?	Road view
5	What is the weather like?	Road view
6	What lane are you in?	Road view
7	What speed are you going at?	Head-up display/instrument cluster
8	What is your EV range?	Instrument cluster
9	What is your fuel range?	Instrument cluster
10	What time is it?	Instrument cluster

Eye tracking was conducted using two BASLER acA640-100gm cameras tasked with measuring head position and eye trajectory (plotted on three axes). Eye-gazes were computed using SMART Eye software (version 7.0) using a construction of a 3D mesh representing the simulator environment and the associated information displays (see Figures 6.2 and 6.3).

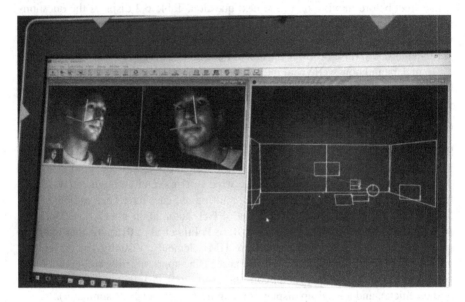

FIGURE 6.2   Eye-tracking interface and areas of interests used for analysis.

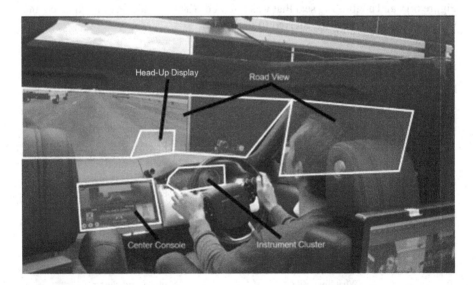

FIGURE 6.3   In-vehicle depiction of areas-of-interest used for analysis.

The driving environment was designed to replicate a three-lane stretch of road with no bends. Sixteen regular road vehicles (i.e., cars, not vehicles like lorries) were generated to surround the vehicle being driven, speeding up and slowing down in line with the main vehicle. This created a traffic 'bubble' in which the participant was part of, so long as they stayed in the middle lane. Weather was set to dry and sunny.

6.2.4 PROCEDURE

On arrival to the laboratory, participants were given a brief verbal introduction to the study and safety aspects. They then read the information sheet and signed a consent and attendance form. Following this, participants completed a demographic questionnaire capturing age, gender, and driving experience both generally and with automated features. Participants were then guided into the driving simulator where they adjusted seat and steering wheel positioning. Drivers were then introduced to the controls and information displays including the cluster information, the head-up display, the central console, audio and verbal interaction, vibrating seat, and ambient lighting. Drivers were then introduced to what will happen in the experiment outlining how and when transitions were expected.

Drivers took part in a shorter TOOTL condition (7 min in total) to become familiar with the vehicle's controls and how to interact with the system. Drivers were then instructed that there were four trials in total, potentially taking up to 35 min to complete, and that breaks are encouraged prior to each trial. When the participant was satisfied that any remaining questions were addressed, the trials began.

For each trial, participants started in the hard shoulder of a motorway and were instructed to drive off into the middle lane and keep to the local speed limit (70 mph). After a minute, the system vocalized to the driver that automation was available and presented the driver with a tone and visual wording/icon accordingly. Participants passed control to the vehicle by pressing two flashing green buttons with their thumbs on the steering wheel. Once activated, the black lighting of the displays and the orange ambient lighting transitioned to white displays/blue lighting. Vocal indications were given throughout such as 'the car has control'.

During automated control, to simulate a secondary task the driver played Tetris on a Window's tablet, and they were told that score was being recorded. The secondary task was implemented to ensure that driver's attention was directed away from the driving task during automated control to measure how drivers raise awareness during the control transition process. A visual indicator counting down the time left in automation (from 1 min or 10 min depending on condition) was displayed on all three screens by default. At 5, 2, and 1 min before manual control was expected, an audio tone and a vocalized alert was given to the driver notifying them of time remaining. When the countdown reached zero, the seat vibrated in co-occurrence with an audio and vocal alert. The handover icon animated the requirement to resume driving position. At this stage, the vehicle vocalized questions and displayed them on each display. Questions asked the driver about vehicle status (e.g., fuel left and speed) or the driving environment (e.g., what color is the vehicle in front). Each answer was delivered vocally and was categorized as being either correct or incorrect by the researcher. Once the car was satisfied that more than half of questions were correctly answered, the vehicle indicated the driver to take control of the vehicle by vocally and visually communicating with the driver. Should questions come below the 50% threshold, an additional warning was given to the driver, but the handover was still initiated. After pressing the two green buttons the driver was now in control; audio, vocal, visual alerts, and ambient lighting (now orange) were given and the vibrating seat pulsed one last time to confirm the handover. This process represented one control cycle and was performed three times for each condition (12 for entire experimental session) before being asked to pull over to the hard shoulder and bring the vehicle to a stop.

Once four trials had been complete, the driver left the vehicle and took part in a final debrief questionnaire where they were able to report their preferences and opinions about the handover procedures they had experienced. They were then thanked for their time, notified of payment, and advised not to drive for another 20 min.

6.2.5 METHOD OF ANALYSIS

The following analyses were conducted to assess the nature of both handover and post-handover (manual driving) visual gaze behavior:

- Repeated measures ANOVA for testing differences in total gaze durations during the vehicle-to-driver handover process between four primary visual streams (road, cluster, head-up display, and center console).
- A Pearson's correlation between the visual streams that were vocally prompted to, and actual visual gaze duration, for testing how well vocal interaction guides visual search.
- How demographic and situational factors affected gaze durations—gender, age, type of vehicle the driver owns, and TOOTL.

6.3 RESULTS

Gaze data were analyzed using R-Studio v.0.99.902. Bar plots were generated using the ggplot2 package. Post-experiment data processing involved trimming the data to isolate the time between the handover alert and the resumption of control from the driver. The four primary visual information streams were used for analysis (road, instrument cluster, head-up display, and center console). Frequencies of counts (1/50th of a second) were transformed into a mean time for a single handover in seconds for each participant and mapped to the associated area of interest. The data were analyzed by testing for overall differences in area of interest gaze times and then with the addition of explanatory variables such as gender, age, class of car owned, and TOOTL.

6.3.1 HANDOVER PROCESS VISUAL GAZE DURATIONS

6.3.1.1 Overall Visual Gaze Duration

Figure 6.4 displays overall differences in total gaze duration between each area of interest, displaying mean seconds for each area. Overall handover time varied based

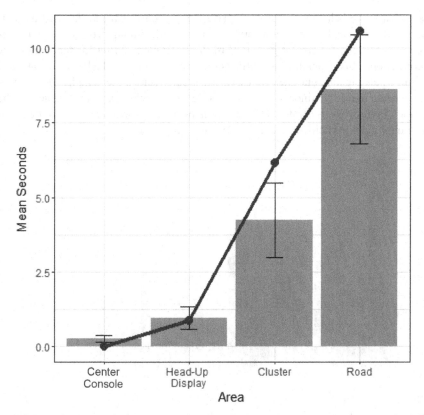

FIGURE 6.4 Overall mean total gaze duration from the notification to the takeover of control displayed. Error bars represent confidence (95%) intervals. Line plot indicates the expected gaze time in relation to vocal prompts.

on a number of factors: how many questions participants selected to answer during the handover period after the first trial (default 5 could be changed to 1–10) and how long drivers took to respond with correct answers to vocal prompts. Finally, gaze times that were unaffiliated to an area of interest were excluded from the plots and analysis. A repeated measures ANOVA found that there was a main effect of area on gaze duration, $F(3, 124) = 72.64, p < .001, \eta p^2 = 0.69$. Pairwise t-tests corrected with the Holm–Bonferroni method showed that there was a significant difference in mean gaze duration for all comparisons of areas of interest ($p < .001$).

Based on the distribution of where answers resided within the visual streams, total gaze duration was tested for a correlation with the percentage of questions that were implemented during handover protocol across the entire study. These distributions were as follows: road view—60%, instrument cluster—35%, head-up display—5%, and center console—0%. There was a significant Pearson's correlation between vocal guidance and associated gaze durations toward the assigned visual stream ($r = 0.71, p < .001$).

Figure 6.5 displays visual gaze times as a stacked bar plot representing everyone who took part in the trials. The stacked bar plot reveals four distinct features of individual gaze behavior. 1) Overall, drivers appeared to gaze at the road as their main form of visual information source, varying little among drivers. 2) There was a consistent trend for greater gaze toward the instrument cluster for supplementary information. This seemed to vary little across participants. 3) The HUD was utilized by most participants, but gaze times varied greatly across the sample. 4) The central console played a minor role in handover gaze behavior, varying little across the sample. Figure 6.6 illustrates these differences as an expression of percentage for each individual participant rather than the absolute value.

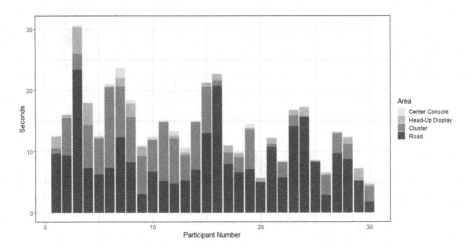

FIGURE 6.5   Handover total gaze duration in seconds toward areas of interest displayed across participant number.

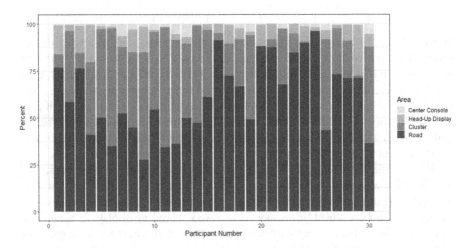

FIGURE 6.6   Handover gaze duration displayed as percentage toward areas of interest displayed across participant number.

6.3.2 Demographics and Behavioral Factors

It was hypothesized that demographical factors such as gender, age, and class of vehicle owned would have an effect on gaze behavior during handover. Mixed-effects ANOVAs found no significant main effect for each variable (gender, age, vehicle type owned, and time out of the loop; $F(1, 115) = 0.44$, $p > .05$, $\eta p^2 = 0.01$; $F(2, 114) = 1.33$, $p > .05$, $\eta p^2 = 0.026$; $F(1, 115) = 1.02$, $p > .05$, $\eta p^2 = 0.0003$; $F(1, 235) = 2.18$, $p > .05$, $\eta p^2 = 0.003$, respectively). Figure 6.7 shows the effect of these factors on handover gaze behavior. The spread of gaze time in these figures show the variability in gaze behavior within groups—no clear differences arise for every category during analysis.

As demographic differences such as gender are typically low, a post-hoc G-power analysis suggests that with effect sizes as low as 0.44, a minimum sample size recommended is within the region of 70 drivers.

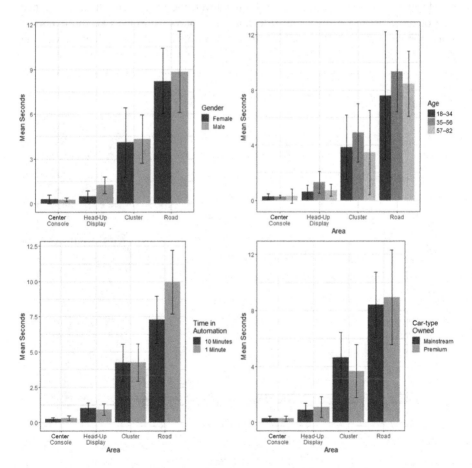

FIGURE 6.7 Handover gaze times displayed for each demographic recorded (age, gender, out-of-the-loop condition, and car type owned).

6.3.3 Post-Handover (Manual Driving) Visual Gaze Durations

Figure 6.8 displays data from 60 s of manual driving following the handover process. It appeared that drivers were primarily gazing at the road with supplementary interface information coming from the head-up display and instrument cluster. A repeated measures ANOVA showed that there was a significant main effect of area on gaze duration, $F(3, 116) = 207$, $p < .001$, $\eta p^2 = 0.84$. Post-hoc pairwise t-tests corrected using the Bonferroni–Holm method showed that there was a significant difference between every comparison of area ($p < .001$) except for the head-up display paired with the center console ($p > .05$).

Analogous to that of Figures 6.5 and 6.6, the stacked bar plots in Figures 6.9 and 6.10 show overall visual gaze toward the four main areas of interest during manual driving for each participant. The data reveal noticeable trends: 1) overall, drivers are heads-up, focusing on the road environments. 2) There was a greater reliance on head-up displays for information when compared to visual gaze behavior during handover interactions. 3) Similar to the handover, the instrument cluster remained one of the most relied on visual information displays. 4) Much like the handover data, the central console played a minor role in handover gaze behavior, varying

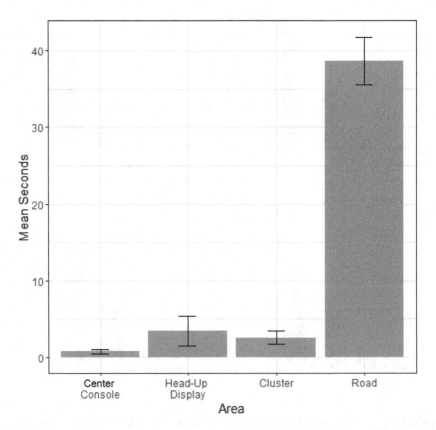

FIGURE 6.8   Overall post-handover mean gaze times displayed with confidence (95%) intervals.

little across the sample. Figure 6.9 represents these differences as raw times, whereas Figure 6.10 illustrates these differences as an expression of percentage rather than absolute value. These percentages illustrate how these proportions vary between drivers.

For demographic differences, mixed-effects ANOVAs found no significant main effect for most variables (age, vehicle type owned, and TOOTL; $F(2, 114) = 0.52$, $p > .05$, $\eta p^2 = 0.009$; $F(1, 115) = 0.21$, $p > .05$, $\eta p^2 = 0.002$; $F(1, 235) = 0.32$, $p > .05$, $\eta p^2 = 0.001$, respectively). A mixed-effects ANOVA for gender differences was significant, $F(1, 115) = 7.05$, $p < .01$, $\eta p^2 = 0.06$, showing that males and females

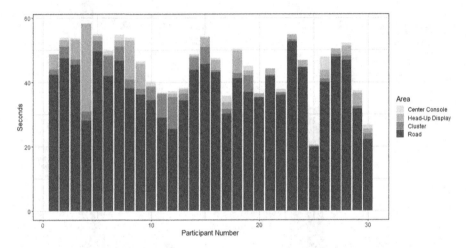

FIGURE 6.9 Post-handover (manual) gaze time in seconds toward areas of interest displayed across participant number.

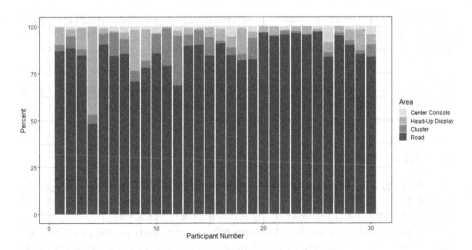

FIGURE 6.10 Post-handover (manual) gaze time in percentage toward areas of interest displayed across participant number.

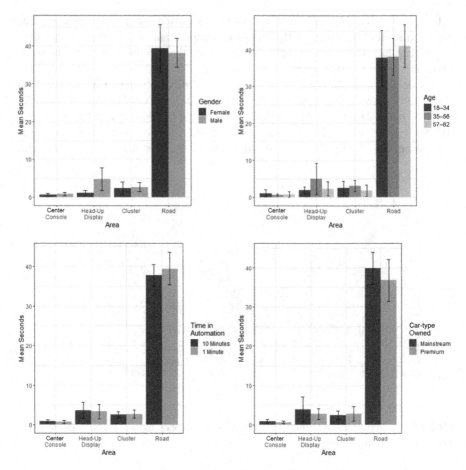

FIGURE 6.11 Post-handover (manual) gaze times displayed for each demographic recorded (age, gender, out-of-the-loop condition, car type owned, and whether drivers selected high or low amounts of visual information to be displayed on interfaces).

differed significantly. Figure 6.11 shows the effect of these factors on handover gaze behavior, indicating once again that there is a great amount of variability within groups, resulting in less prevalent group differences.

6.4 DISCUSSION

This study explored the nature of visual gaze durations toward differing visual information streams during handovers from a conditionally/highly automated vehicle (C/HAV) to the driver and manual driving following these handovers.

Overall, during the experiment drivers relied significantly on the road environment as their primary source of visual information, indicating that although HMIs are important for the handover process (Eriksson & Stanton, 2018; Walch et al., 2015), it must not be forgotten that the real-time road environment is where drivers will look to gain accurate and detailed information. It follows that if the driver is to focus on the road environment, handover assistants could be implemented in a way to guide or supplement visual gaze toward important cues rather than rely on head-down displays alone (e.g., Tesla, 2018). This could be implemented in tandem with vocal interaction (Clark et al., 2018; Eriksson & Stanton, 2017b; Large et al., 2017) or via more explicit head-up displays indicating road hazards through augmented reality (Eriksson et al., 2019).

Perhaps the most salient finding from this experiment was the high correlation between audio guidance toward areas of interest and visual gaze behavior. This indicates that vocal guidance could serve as a robust tool in guiding drivers' attention to relevant pieces of information for the purpose of transactions in situation awareness (Sorensen & Stanton, 2016), advancing on previous findings in eye-gaze behavior during handover interactions (Clark et al., 2018) indicating that designing vocal interactions to promote visual guidance should be considered by automotive designers and manufacturers alike.

During handover interactions the instrument cluster, a short distance below the road view, was the visual HMI component that had the statistically significant longest gaze duration out of all visual streams, for the purpose of handover interaction. Instrument clusters are more populous among road-vehicle models than both head-up displays and central consoles, which are more likely to be found in prestige vehicles. The technology acceptance model (TAM3; Venkatesh & Bala, 2008) provides a way of being able to explore these findings in light of how users uptake new types of technology. The TAM3 outlines a combination of user, system, device relevance, perceptions, and social interactions that lead to the use or disuse of technology. Within the model, experience with other devices can act as a barrier to uptake, for example, if the user finds that the current technology is easier to use due to experience, even if it is less optimal. It follows that drivers may have interacted with the instrument cluster as this is a component that they are more familiar with compared to the other visual displays. This may lead to an increased perceived usefulness and, in turn, intention to use. The cluster display also requires less physical effort to glance at if the driver is in a standard driving position and therefore may be preferred over the center console.

The TAM3 may also explain why there was a great amount of individual difference in the gaze duration toward the head-up display during handover and manual interactions. It may be that those with current HUD technology in their vehicles are more likely to use this visual information stream as their experience may interact with their perceived usefulness and intention to use. In this experiment, the head-up display was limited to a small subsection, just above the view of the steering wheel. It was not addressed whether an entire screen augmentation would better guide a driver's search for visual information.

One recurrent finding during this experiment was the low reliance on the central console compared to the other visual streams made available. The role of the central console in HMI design is contentious, as it is possible to display a lot of information on these screens, such as those found within infotainment systems in current level 3/4 models and concept designs (e.g., Tesla S Class and Audi A8; Tesla, 2018; Audi, 2019a), but little is known about their practical affordances. The nature of the central console during handover interactions has not yet been established, although our findings show that the role for handover information on this particular display may not be as important as those that are found within the road environment or screens closer to this area.

For post-handover manual driving, drivers' attention for the most part was heads-up. Minor glances toward each display were made, with some drivers opting to rely on the head-up display for supplementary information.

It was hypothesized that certain demographics of drivers may have differing gaze duration behavior when it comes to the handover task (Feldhütter et al., 2017; Sun et al., 2016; Yan et al., 2016). Our findings show that, visual gaze toward visual displays did not differ with respect to demographics such as gender, age, and whether the driver drove a premium or mainstream vehicle. These findings extend to other such variables such as time out of the loop and how much information was kept on visual displays. However, there was a significant effect for gender on the reliance on head-up displays, indicating that males may favor this visual stream for manual driving more so than females. Previous findings show that visual behavior regarding scan paths and vigilance are variables that differ (Feldhütter et al., 2017; Sun et al., 2016; Yan et al., 2016); however, our findings indicate that this may not extend to which visual streams drivers utilize during handover interactions. It follows that perhaps providing a default option for drivers that cover road-based guidance and cluster information, as well as high amounts of HUD customizability (due to the high within-subjects variance in gaze duration), might be the best approach to cater for a broad range of driver characteristics.

This experiment contributes toward how the role of directability (Clark, 1996; Klein et al., 2004, 2005) can be implemented as a method of vehicle-to-driver communication. Based on these findings, directing attention toward relevant visual information as well as focusing on head-up appear to be valid as a way of implementing this approach. It follows that, based on our road gaze time duration findings, guiding the driver to important areas within the environment may allow for drivers to raise their own SA during transactions (Sorensen & Stanton, 2016; Stanton et al., 2006; Stanton et al., 2017b). This brings into question how modern AV interfaces are being designed, as many manufacturers implement a center console approach for interacting with the driver during TOOTL (e.g., Tesla S-Class; Tesla, 2018, 2020). Our findings suggest that relying on such interfaces may be appropriate during automated periods, although a shift of focus toward head-up methods may be more suitable during SA transactions during handover.

Additionally, the variation in eye-gaze behavior between drivers (see Figures 6.5, 6.6, 6.9, and 6.10) shows the requirement for these transactions to be customizable. Manufacturers should remain focused on raising SA for safety but appropriately implement concepts such as directability to do so. In other words, as seen by these findings, designs should ensure that handover assistants are utilized by the driver in a way that they feel is accessible and suits their schema requirements (Neisser, 1976; Stanton et al., 2006).

6.4.1 Conclusions

This experiment measured driver eye-gazes during handover and manual interactions in a level 3 automated vehicle motorway simulation. Eye-gaze durations showed that vocal guidance may serve as an effective tool in guiding visual attention. Further, great variability in the utilization of visual information streams suggests the requirement for customization. Finally, particularly during manual driving following handover, drivers gazed minimally at the center console—suggesting that current AV designs are not addressing the unique requirements of handover interaction and how these visual streams are utilized when coming back into the loop. These findings show how the concept of directability can be implemented into AVs as a means of facilitating situation awareness transactions. Further work should address the ways in which head-up augmentations can be used to guide visual awareness and explore which vocal cues may be most suitable in raising SA during control transitions.

6.4.2 Future Directions

So far, this book has scoped communication theory, the current state of C/HAV HMI, domain constraints, shift-handover practices, how vocal communication can be implemented in human–automation interaction, and how visual information can supplement vocal information. The next stage in the book is to design a testable prototype for C/HAV interaction. Chapter 7 recruits drivers belonging to different skill groups and conducts focus groups, asking users what they suggest for C/HAV interaction during a journey. The chapter brings the book one step closer to a design outcome and demonstrates how user participation can be conducted for human factors design.

Section III

Designing New Interfaces and Interactions for Automated Vehicle Communication

7 Participatory Workshops for Designing Interactions in Automated Vehicles

7.1 INTRODUCTION

This chapter, along with Chapter 8, draws together all previous chapters to generate and design a prototype that addresses the concepts outlined in previous chapters. So far, the following tenets can be assumed:

- Interaction should consider the entire control cycle, not just handover (Chapter 3—cognitive work analysis).
- Information transfer should be bidirectional, situation specific, and give the user control of what information is displayed and relayed (Chapters 4 and 5).
- Allowing the driver to ask questions during the journey appears to be an effective situation awareness (SA) raising strategy (Chapters 4 and 5).
- Automation should communicate its current state, intentions, and perceptions (Chapters 2 and 3).
- Vocal and visual communication together can serve as effective modalities for driver–automation interaction (Chapters 3 and 6).

This chapter presents findings from participatory workshops where the user group (i.e., drivers) is brought together into groups to discuss potential solutions to conditionally and highly automated vehicles (C/HAV) interaction. These focus groups are performed for each skill group, as each skill group will hold different requirements for raising SA.

7.1.1 DRIVER SKILL IN C/HAVS

Distributed situation awareness (DSA) refers to Neisser's perceptual cycle model (Neisser, 1976) to provide insights into how individual agents integrate with their environment, both perceptually and behaviorally. The model outlines that individuals make use of 'schemata'—mental templates constructed from experience that are accessed to interpret environmental cues and guides appropriate behavioral responses. In driving tasks, the idea of every driver possessing unique schemata as a result of past experiential and behavioral events raises an interesting question— How do drivers of differing skill view the handover task in C/HAVs? To begin

DOI: 10.1201/9781003213963-10

answering this question, it is necessary to evaluate previous research on driver skill and behavior.

The relationship between the ability to raise individual SA and driver skill shows that advanced driver training may improve driver SA (Walker et al., 2009). Walker et al. found that advanced driver training may improve working memory capabilities that result in the improvement of driver SA (Walker et al., 2009). Advanced drivers have also been found to pay closer attention to environmental feedback (Bainbridge, 1978; Stanton et al., 2007). This addresses Neisser's perceptual cycle model (Neisser, 1976) as driver behavior (attention) is guided by driver schemata (as a result of experience) and becomes more sensitive to environmental cues. For this reason, it could be predicted that advanced drivers may not require comprehensive handover assistants, as they are more capable of processing environmental information and raise their own individual SA without the need for complimentary interfaces. This is further illustrated by findings presented by Young and Stanton who found that there was an effect of driver skill on the issue of workload in automated driving (Young & Stanton, 2002a). Young and Stanton (2002a, 2002b, 2007b) summarize their findings with reference to the malleable attentional resources theory as an illustration of the inverse relationship between driver skill and mental workload (Gopher & Kimchi, 1989). As skill increases in the driving task, so does automaticity. However, automaticity can lead to a greater level of complacency (Charlton & Starkey, 2011). For this reason, advanced drivers may in part be more capable of performing but may not rely on assistance systems in the process. Further, research into new drivers and hazard perception shows a range of differences to their more experienced counterparts such as less efficient gaze behavior, less vigilant regarding their mirrors, and focus closer to the front of the vehicle (Quimby et al., 1986). It may be that drivers of less experience will benefit from handover awareness assist more due to their attentional shortcomings as learners.

Driver experience and driver age are inherently correlated. Arguably, with age comes changes in the way the driving task is performed, and studies looking into driver skill are inevitably confounded by this variable, a factor recognized across driving research (e.g., Evans, 2004). In controlled experiments exploring takeover performance and age Körber et al. (2016) found little differences in behavioral performance when drivers were asked to take control with and without a secondary task. Morgan et al. (2016) argue that most handover studies have focused on mainly middle-aged and high-mileage drivers. As interaction with automation is dependent on driver skill (Young & Stanton, 2007a) it is important to consider the future of C/HAV operation for the range of driver skill groups. In the current state, C/HAVs will be implemented in vehicles with high prestige. However, over time, C/HAVs will become more accessible to a wide range of drivers with differing skills and experience. It may be that advanced drivers, equipped with their experience, will be more comfortable taking control from a C/HAV as they are familiar with the nature of the roadway; however, a learner would have had less exposure such events.

7.1.2 Current Study and Research Questions

Participatory design is an approach to a design problem that seeks to include the end user in the design process, thus addressing their needs and concerns at the beginning of the design lifecycle (Sanders, 2003). It stems from the philosophy that users have valuable insights that can lead to creative concepts when given the chance to share their ideas (Sanders, 2003). The challenge arises for designers to take user experience and participatory designs and implement them with practical, realistic, and balanced approach (Spinuzzi, 2011). A good illustration of this concept is the implementation of customization and personalization in designs (Bonett, 2001; Brennan & Adelman, 2008). Customization, by definition, allows users to individually tailor preferences/requirements to their own individual needs (The Oxford English Dictionary, 2018a) but should lie within scope of design specifications and legislature. On the other hand, personalization relates to the profiling and creation of a design that can be applied to a target group (The Oxford English Dictionary, 2018b). Both approaches appear to have individual merit; however, understanding how both can be implemented requires a sufficient amount of knowledge about user needs and requirements (Spinuzzi, 2011).

The domain of C/HAV handover lends itself to the participatory design approach due to the availability of a range of demographics to researchers as drivers are part of the general public. This study implements the participatory design approach by asking groups of drivers to discuss and design a C/HAV handover/handback assistant. The following research question was central to the discussions: how do drivers of different skill categories view solutions to the handover and handback problem in C/HAVs?

7.2 METHOD

7.2.1 PARTICIPANTS

Ethical approval was gained via the University of Southampton ethics panel (ERGO II No. 40182). Drivers were recruited across three separate categories as defined by Young and Stanton (2007a): 1) learner drivers—currently having lessons, 2) intermediate drivers—have held a full license for over 1 year, and 3) advanced drivers—have completed an advanced driving course under the Institute of Advanced Motorists (IAM, 2016) or The Royal Society for the Prevention of Accidents (RoSPA, 2018). Due to availability and population demographics, balancing age and gender was challenging (e.g., learner drivers more likely to be young and advanced drivers likely to be older males). These confounding variables are well noted across driving research (Evans, 2004).

Focus groups consisted of between five and seven participants, conforming with size of focus groups outlined by Krueger and Casey (2002, 2015) Participant data are displayed in Tables 7.1 and 7.2 displaying demographics and driving experience.

7.2.2 DESIGN

Focus groups lasted approximately 1.5 h in line with common practice (e.g., HSE, 2003). Driver skill was implemented as a between-group, three-level independent variable categorized into groups of learner, intermediate, and advanced drivers. Additional variables measured from the sessions were age, gender, annual mileage,

TABLE 7.1
Participant Demographics

Skill Category	N	Gender (M/F)	Mean Age
Learner	7	4/3	26.1 (SD = 6.4)
Intermediate	6	1/5	52.6 (SD = 16.8)
Advanced	5	3/2	63.3 (SD = 7.2)

TABLE 7.2
Participant Driving Experience

Skill Category	Mean Annual Mileage	Mean Years Held License	Mean Learner Hours	Mean Years As Advanced Driver
Learner	—	—	52.6 (SD = 39.7)	—
Intermediate	6,500 (SD = 3,500)	32 (SD = 15.6)	—	—
Advanced	12,600 (SD = 2,191)	47 (SD = 9.4)	—	18.3 (SD = 10.6)

and years holding license (and where applicable, hours of lessons and years of holding advanced qualification). Dependent variables from this study were the output from audio data and collective written design suggestions using post-it notes on a predesigned template.

7.2.3 Procedure

Participants were recruited either via the University of Southampton's internal webpage (targeting learners/intermediates), through driving schools/instructors (targeting learners), or through contacting IAM or RoSPA (targeting advanced). Prior to taking part in the focus group, each participant filled out a demographics questionnaire. An introduction was given by the first author to inform participants of the current state of C/HAVs, the challenges faced by designers (i.e., SA, usability, and trust), and the interface elements that are available to designers in current C/HAVs such as head-up and head-down displays, vocal assistants, audio cues, haptic feedback, and inputs. Following this, groups were then introduced to one of three storyboard scenarios. In scenario 1, participants were asked to discuss and generate design concepts related to 1-h out of the loop (OOTL). Scenario 2 followed a similar story line with a change to 10 min. Scenario 3 represented a general scenario with no time constraints asking what participants would like from the handback procedure (interfaces prior to activating automation).

Participants discussed their thoughts in relation to their driving experiences and a range of factors that they could consider. These factors were drawn from the cognitive work analysis within Chapter 3 representing the different modalities available (e.g., audio and visual), the types of visual displays (e.g., head-up/head-down display), and timings from the social organization and cooperation analysis—contextual activity template (SOCA-CAT). This framework was presented to the group in the form of a diagram representing stages and swim lanes. They were asked to generate information types and how they are delivered as a group, using their collective experiences and wrote onto a blank swim-lane diagram placed in the center of the room. The facilitator's role during discussions was to ensure that timings were met and write up the group's ideas on post-it notes as well as prompt participants as to what information stream they desired each process to be allocated to, if not already specified.

7.2.4 METHOD OF ANALYSIS

The data from this study were derived from demographic information, design concepts, and audio recordings. For each suggestion that the group agreed upon (as derived from audio recordings and written task), a schematic was digitally generated for each group as outlined in their written design concepts. A thematic analysis of transcripts was conducted to determine what aspects of handover protocol each group focused on during discussions—attributions made by two analysts of a random selection of suggestions (comprising 40% of total suggestions) agreed on 83.3% of statement-theme parings, exceeding the widely accepted minimum requirement for agreement (McHugh, 2012).

7.3 RESULTS

Figure 7.1 displays the frequency of design suggestions made by drivers during focus groups regarding the handover (driver-to-vehicle) in relation to five main themes. Themes were generated by the grouping of each suggestion (learner = 18, intermediate = 14, and advanced = 16):

- Alert—Any suggestion made that relates to the driver being made aware that a handover is required (e.g., audio tone and flashing light on dashboard).
- Check Arousal—Any suggestion made that relates to the vehicle assessing the physical and/or cognitive arousal of the driver, whether that be through requests to actively respond or implicit methods such as built in eye tracking.
- Choreography—Any suggestion made that relates to timings, coordination, or clarity of the handover process. Examples include the state of automation and the time left until handover is expected.
- Aid Awareness—Any suggestion made that indicates a requirement for the vehicle to feed information to the driver related to the past, present, or future state of the driving scenario. Examples include hazards and route planning information.
- Transition—Any suggestion made that relates to the physical aspect of taking over control of the vehicle (e.g., taking control of pedals before steering wheel).

The stacked bar plot (Figure 7.1) shows that discussions in the learner group focused more on alerts and awareness assist, whereas as driver experience increased, a shift away from awareness assist toward designing better and smoother ways of transition-ing control was observed—this is better presented in Figure 7.2. To understand these differences better, each group's final designs and comments are discussed in detail in the following sections.

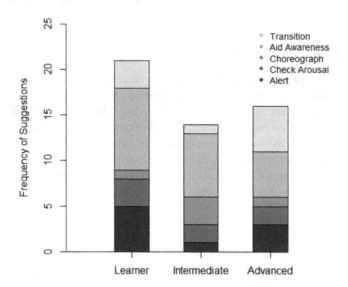

FIGURE 7.1 Summary of themes generated from focus group transcripts, split across skill groups.

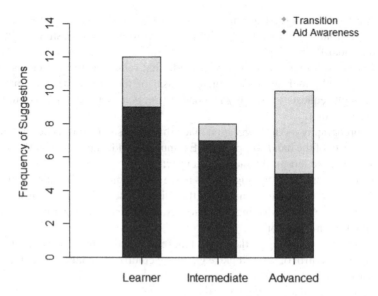

FIGURE 7.2 Frequency of suggestions comparing the themes 'aid awareness' and 'transi-tion' between driver skill groups.

7.3.1 LEARNER HANDOVER DESIGN

Figure 7.3 displays the design schematic created by the learner drivers' group. Overall, this group preferred visual interfaces to guide handover interactions. Starting with a vibrating seat '3 to 5 minutes before takeover was expected', notifications would steadily increase as the takeover time drew nearer. Vocal and audio notifications (e.g., polite tone or vocal instruction) would reduce radio volume and interact with electronic devices that may be in use (e.g., smartphone). The group thought it would be important for the alerts and timings to be specific to the secondary task that was being performed (e.g., emailing, reading, and conversing). From here, the driver provides an input (e.g., squeezable portion of the wheel or sensors on the wheel) to verify they are aware that they are required to take control.

During the information stage, the group thought it would be important to receive real-time information about where they were and what the situation was like up ahead. GPS info was expected to be displayed on a center console, with augmented traffic information on the HUD. Customizable vocal output was also desired at this stage. Concurrently, the group wanted information such as current gear and current speed displayed on the HUD and any dangerous weather information to be displayed both on the head-down display (HDD) and HUD. From here, drivers wanted a way to raise awareness of what was around them through the use of surround cameras and blind spot indicators. Finally, prior to the takeover process, the group requested a way to reset driving position (seat and wheel position) to that of driving rather than secondary task; however, in a level 3 vehicle this may not be feasible due to the potential for an emergency takeover.

To take control, the learner focus group wanted an input in button form, due to its reliance, and a clear signal that control is now in the hands of the driver. Further, this group requested that automation would oversee their driving performance following handover. Throughout both automated and manual control, drivers requested that the HDD showed when it was in automated mode, a clear color tint of the display (e.g., amber for manual and green for automated). Finally, while in automated mode, this group requested there to be a timer present on the HUD indicating when driver control was expected and a manual override capability.

For shorter journeys, the group found it to be important to customize the amount of time that is given for alerts to be triggered. It was widely thought that for a 10-min automation period, 1 min was sufficient to be alerted, raise SA, and regain control. Regarding interfaces, the group decided that a lot can change in 10 min and kept the same design as the 1-h out-of-the-loop scenario.

7.3.2 INTERMEDIATE HANDOVER DESIGN

Figure 7.4 displays the design schematic created by the intermediate drivers' group. Overall, this group preferred visual interfaces in tandem with a vocal interface to guide handover interactions. The group discussed how they want the vehicle to have a typical cluster (speedo, RPM, fuel, and engine temperature). Starting with the radio volume being lowered, the group requested a 'nonthreatening' jingle or tone to be presented, which increased in intensity over time. Next, they wanted information

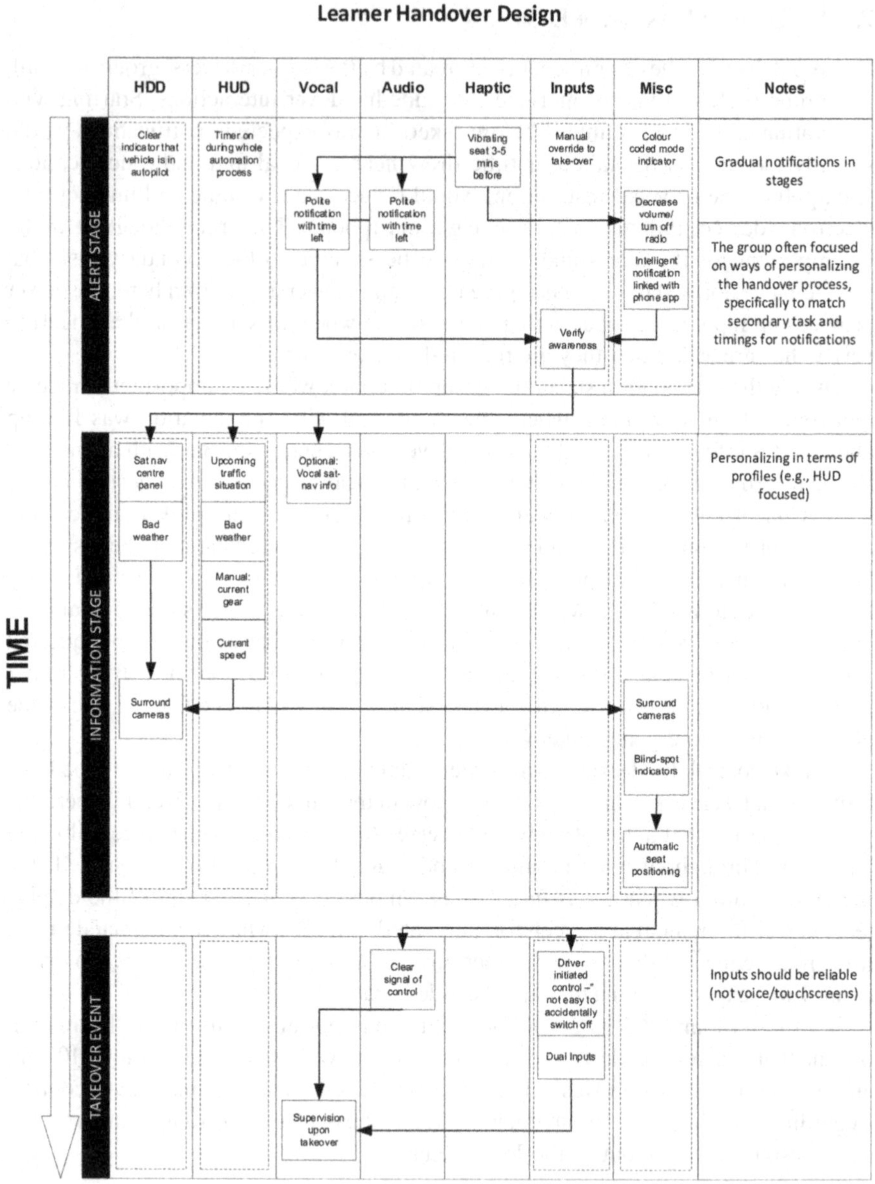

Learner Handover Design

	HDD	HUD	Vocal	Audio	Haptic	Inputs	Misc	Notes
ALERT STAGE	Clear indicator that vehicle is in autopilot	Timer on during whole automation process	Polite notification with time left	Polite notification with time left	Vibrating seat 3-5 mins before	Manual override to take-over; Verify awareness	Colour coded mode indicator; Decrease volume/turn off radio; Intelligent notification linked with phone app	Gradual notifications in stages / The group often focused on ways of personalizing the handover process, specifically to match secondary task and timings for notifications
INFORMATION STAGE	Sat nav centre panel; Bad weather; Surround cameras	Upcoming traffic situation; Bad weather; Manual: current gear; Current speed	Optional: Vocal sat-nav info				Surround cameras; Blind-spot indicators; Automatic seat positioning	Personalizing in terms of profiles (e.g., HUD focused)
TAKEOVER EVENT			Supervision upon takeover		Clear signal of control	Driver initiated control –" not easy to accidentally switch off; Dual Inputs		Inputs should be reliable (not voice/touchscreens)

TIME

FIGURE 7.3 Handover design from the learner driver group. Lanes indicate modality and nodes indicate type of information. Vertically connected/adjacent nodes indicate concurrent presentation. Y-axis indicates time. Dotted boxes indicate that the node applies throughout.

on the HDD, HUD, and on vocal to notify them of how long they have left, customized to 5 or 10 min prior to takeover. In tandem, the group requested distance left until takeover was expected, alongside any route, journey, and traffic information displayed both on the HDD and vocally.

Intermediate Handover Design

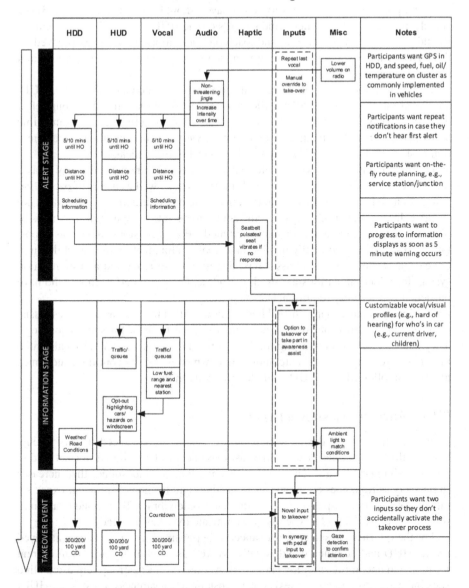

	HDD	HUD	Vocal	Audio	Haptic	Inputs	Misc	Notes
ALERT STAGE				Non-threatening jingle		Repeat last vocal / Manual override to take-over	Lower volume on radio	Participants want GPS in HDD, and speed, fuel, oil/temperature on cluster as commonly implemented in vehicles
				Increase intensity over time				Participants want repeat notifications in case they don't hear first alert
	5/10 mins until HO	5/10 mins until HO	5/10 mins until HO					
	Distance until HO	Distance until HO	Distance until HO					Participants want on-the-fly route planning, e.g., service station/junction
	Scheduling information		Scheduling information					
					Seatbelt pulsates/seat vibrates if no response			Participants want to progress to information displays as soon as 5 minute warning occurs
INFORMATION STAGE						Option to takeover or take part in awareness assist		Customizable vocal/visual profiles (e.g., hard of hearing) for who's in car (e.g., current driver, children)
			Traffic/queues	Traffic/queues				
			Low fuel: range and nearest station					
		Opt-out highlighting cars/hazards on windscreen						
	Weather/Road Conditions						Ambient light to match conditions	
TAKEOVER EVENT			Countdown			Novel input to takeover		Participants want two inputs so they don't accidentally activate the takeover process
	300/200/100 yard CD	300/200/100 yard CD	300/200/100 yard CD			In synergy with pedal input to takeover	Gaze detection to confirm attention	

FIGURE 7.4 Handover design from the intermediate driver group. Lanes indicate modality and nodes indicate type of information. Vertically connected/adjacent nodes indicate concurrent presentation. Y-axis indicates time. Dotted boxes indicate that the node continues throughout the swim-lane alongside other nodes.

For the information stage, the group decided on having the different input options to either override (directly to takeover control) or receive awareness information. At this stage, if the driver has yet to respond, the seatbelt will pulse and the seat vibrates. The group then requested the following: 1) upcoming traffic or queues at the junction to be displayed both on the HUD and vocally; 2) concurrently, drivers wanted fuel state and potential refill stations to be displayed should it be of importance; 3) the vehicle highlights other vehicles and hazards on the HUD that the driver can cancel should they wish; and 4) weather and road condition information to be displayed on the HDD as well as ambient lighting to match (e.g., green for good conditions and red for dangerous conditions).

To takeover, the group felt it necessary to have a countdown on the HDD, HUD, and vocal interfaces in a 300/200/100-yard fashion. At this point, the group felt it necessary to have two inputs on the steering wheel that are activated at the same time to avoid accidental deactivation of autopilot, as well as pedal input and gaze detection to show that the driver is ready to take control of the vehicle.

During the entire process, the intermediate focus group requested a way of asking the vehicle to repeat the last piece of information, as well as a way of overriding automation and taking control instantly. Further, to make sure the handover does not surprise drivers, the group requested the ability to be able to input new directions or plans to the vehicle while it is in automated mode. Finally, customization was important for this group, specifically in relation to confounding factors such as children in the car interfering with vocal assistants; this group also requested a customization system for which modalities are used, as well as being tailored to those who have particular accessibility requirements.

The intermediate group believed it to be unimportant to have any awareness assist or dangerous weather and road conditions displayed to them for shorter journeys. They suggested a 2-min warning for 10 min in automation—potentially calculated based on percentage of total automation time, which is vocally communicated once and is then followed by a HUD countdown display in time.

7.3.3 ADVANCED HANDOVER DESIGN

Figure 7.5 displays the design schematic created by the advanced drivers' group. Overall, this group preferred smooth transitions of control with little information displayed with regards to awareness assist. Starting with a 'quiet sound' that increases in intensity as time goes on, alongside a flashing light to show that the car expects a takeover soon. If no response is given, the seat would vibrate. To respond to automation, this group preferred to vocally communicate that they are ready.

For the information stage, the advanced group wanted ice information displayed on the HUD and communicated vocally, as well as vocal indication as to current location and upcoming situation (junction). At this stage, this group requested blind spot indicators to be active. Next, opt in customized awareness assist (e.g., HUD/ vocal) was suggested but not forced upon the driver.

To make the transition, this group suggested that the driver could drive the vehicle for a specific amount of time, without actually having control; once this was done for a set amount of time, the vehicle would know that the driver knows what they are doing and would transition control with a vocal countdown (i.e., 1 mile, ½ mile, 3, 2, 1).

Just like the other groups, the advanced driver group made it clear that they would like a manual override in the form of dual inputs, as well as customizable profiles for handover assist suited to the driver and the secondary task to be performed.

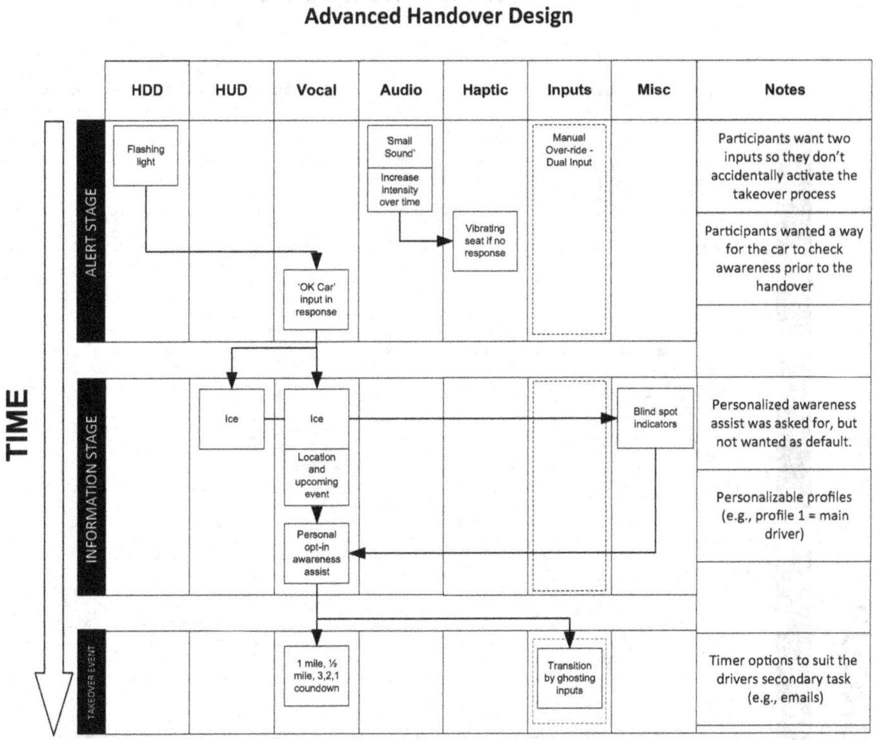

Advanced Handover Design

	HDD	HUD	Vocal	Audio	Haptic	Inputs	Misc	Notes
ALERT STAGE	Flashing light			'Small Sound' Increase intensity over time	Vibrating seat if no response	Manual Over-ride - Dual Input		Participants want two inputs so they don't accidentally activate the takeover process
			'OK Car' input in response					Participants wanted a way for the car to check awareness prior to the handover
INFORMATION STAGE	Ice		Ice Location and upcoming event Personal opt-in awareness assist				Blind spot indicators	Personalized awareness assist was asked for, but not wanted as default.
								Personalizable profiles (e.g., profile 1 = main driver)
TAKEOVER EVENT			1 mile, ½ mile, 3,2,1 coundown			Transition by ghosting inputs		Timer options to suit the drivers secondary task (e.g., emails)

TIME

FIGURE 7.5 Handover design from the advanced driver group. Lanes indicate modality and nodes indicate type of information. Vertically connected/adjacent nodes indicate concurrent presentation. Y-axis indicates time. Dotted boxes indicate that the node applies throughout the entire handover procedure.

There was little comment by the advanced driver focus group as to how they had change the handover process given the amount of time out of automation. However, they made it clear that they require an instant takeover input and that the process should not be 'too long'.

7.3.4 HANDBACK DESIGNS

7.3.4.1 Learner Group

Figure 7.6 displays the handback designs outlining how groups approached the action of transferring control back to automated driving when it is available. Generally, handback designs considered driver trust with the automated system and the physical action of transferring control. Groups differed in their approaches. The learner group preferred to have a universal audio tone displayed and the amount of time of automation available displayed in the HUD. Concurrently, they requested pages in an HDD console to show route info, the awareness the car has of the environment, and the car's future actions so that 'faith' can be built up prior to activation. Next, a

Collated Handback Designs

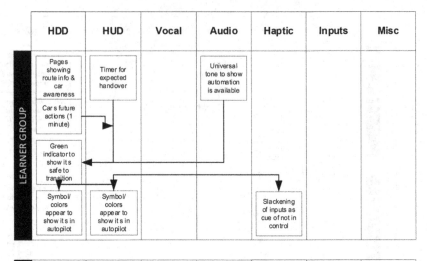

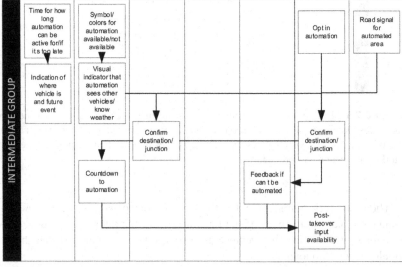

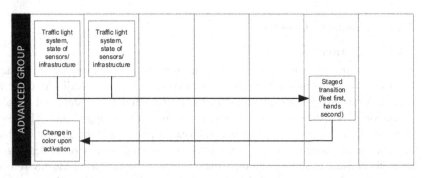

FIGURE 7.6 Handback designs for each driver group. Lanes indicate modality and nodes indicate type of information. Vertically connected/adjacent nodes indicate concurrent presentation. Y-axis indicates group.

green indicator is to be displayed on the HDD indicating a safe transition, and finally, symbols and colors are to be displayed across HDD and HUD to show the transition of control to automation as well as the slackening of controls.

7.3.4.2 Intermediate Group

The intermediate group suggested a more interactive handback design involving a range of driver inputs. To begin, this group desired physical road signs indicating that automation can be safely activated within a certain region. In the vehicle they wanted the amount of time that automation can be active to be displayed on the HDD alongside colors/symbols on the HUD indicating that automation is available. The group made it clear that they wanted automation to support an opt in approach. Following this, this group requested for the HDD to display information showing that the vehicle knew where it was and what it was going to do in the future (e.g., turn-off/handover). Concurrently, on the HUD, visual indicators that the vehicle is aware off, and processing the movements of other vehicles, as well as weather conditions, were requested. Before giving control, this group requested the option to confirm destination and junction through vocal interaction with the vehicle (input and feedback). Following the input of taking control, this group requested a countdown before control is handed back and feedback through vibrations if the automation is not able to be activated. When control is given to automation, this group wanted communication to continue through the use of button inputs.

7.3.4.3 Advanced Group

The advanced group favored a simple approach to the handback—a clear indication of whether it is safe to activate represented using a traffic light system (representing sensors and infrastructure status) on the HDD and HUD. They requested for the transition to automated control to be staged starting with pedals, and then the steering wheel, to gradually build trust. Finally, a change in color across the HDD indicates that automation is now in control.

7.4 DISCUSSION

This study explored how drivers of differing experiences envisaged handover assistants with the intention of promoting safety while being usable and calibrated to suitable levels of trust. This was done through the use of focus groups to explore in detail how drivers viewed the issue of handover while being TOOTL for differing periods of time (1 h vs. 10 min) as well as how control should be handed back to automation.

7.4.1 COMPARISON OF GROUPS' HANDOVER DESIGNS

As the intermediate and advanced group were similar in mean age (52.5 and 63.3, respectively), comparisons can be made more easily between focus group designs. Overall, the advanced driver group expressed less reliance on the HMI for awareness and strategic planning and focused more on the transition itself. By way of contrast, learner and intermediate drivers expressed a preference for HDD, HUD, and vocal interactions to raise awareness and guide the handover. As driver SA has been shown to be related to advanced driver training (Stanton et al., 2006; Walker et al., 2009) it seems plausible that advanced drivers expressed a greater willingness to detect and process environmental cues themselves (Stanton et al., 2006; Bainbridge, 1978) without the assist of a handover human–machine interface (HMI) (e.g., 'Keeping my eye on the road at a certain time . . . so if I need information, it'd have to be projected onto a windscreen—but I'm not sure I need all of that information', advanced group, participant 2, lines 239–242).

As automaticity and workload are related to driver skill (Young & Stanton, 2007a) learner and intermediate drivers may be more reliant on multiple streams of information such as concurrent displays for handover notifications when compared to their advanced counterparts (learners—traffic/location data on HDD, HUD, and vocal; intermediates—notification and route data displayed on HDD, HUD, and vocal; e.g., 'maybe you have something . . . you feel something, then you see something and then you hear something'; learner group, participant 5, lines 22–25). This is supported by research showing a reduction in perceived mental workload and greater user preference with the presentation of multimodal feedback (Oviatt, 1997; Vitense et al., 2003). This finding also supports previously proposed handover HMIs and the effectiveness of multimodal vehicle-initiated takeover requests (Walch et al., 2017; Petermeijer et al., 2017a, 2017b; Politis et al., 2014).

Perhaps the most discussed topic across all groups was the need for customization; a topic previously explored by Bazilinskyy and his colleagues (Bazilinskyy & de Winter, 2015; Bazilinskyy et al., 2015). The concept of customization in this study centered primarily on the modality in which information is displayed to the driver (e.g., 'Well, I know I've got two kids in the back, and they're going to be singing or something, . . . So vocal things aren't going to cut it today'; intermediate group, participant 4, lines 600–602) but also how much time in advance drivers were alerted for the purpose of concluding secondary tasks or time out of the loop (TOOTL) (e.g., 'I know I'm waiting on an important email, no matter what time you wake me up, I'm going to have to somehow finish it, and slam down my laptop, so I want 2 miles notice, not one'; advanced group, participant 3, lines 315–317; 'you're going to Romsey (from Southampton) you set the timer to 1 minute, and if you go to Newcastle you set it to 3 to 5 minutes'; learner group, participant 5, lines 340–343). However, recommendation of what default, safety-critical settings should be was communicated by the advanced group (e.g., 'Well, then the default state needs to be safety'; participant 4, advanced group, lines 453–454). Solutions across the groups involved being able to make changes to the handover protocol prior to setting off for a journey in the form of a system that allows for preset and customizable profiles (e.g., 'Well, maybe there should be different modes. One that has the voice, one that has haptics, and then one where it has haptics and lights, and written stuff'; learner group, participant 6, lines 126–127).

All groups requested a way for the transition to require more than one input to avoid accidental deactivation, for example, the intermediate group suggested two sensors on the steering wheel in concurrence with detection of pedal inputs before automation can be deactivated. All groups requested information about where they were and what was coming up, either in the form of HUD augmentation coming up to junction or as a regular sat-nav implemented on the console.

7.4.2 Changes When Shorter Time Out of the Loop

There has been little research regarding TOOTL and the changes to be made regarding its effect on handover performance from automation to human operator. Typically, the energy production domain favors allocating more time to handover if the incoming staff have been absent for a longer period of time (Adamson et al., 1999; Lewis & Swaim, 1988). There is a clear need in this domain to commit more resources to the handover task when TOOTL is higher. This study explored this concept in the automated vehicle (AV) handover task and asked what drivers may want to change in a handover HMI in the scenario of 10 min vs. 1 h.

Most of the discussion regarding TOOTL in both the learner and the intermediate group surrounded the amount of time prior to the takeover that alerts were given. Learners requested shorter journeys to give a 1-min warning (e.g., 'I think the timings could be shorter, if it's only 10 minutes you don't need a 3-minute warning'; learner group, participant 3, line 327), whereas intermediates preferred a more sophisticated system that calculates a percentage of the TOOTL for the amount of time given prior to takeover (e.g., 'Percentage based on time overall for planned automation'; intermediate group, participant 3, line 551). The learner group kept their information streams the same, whereas intermediates stated that it was not important to have either awareness assist or information about weather/road conditions (e.g., 'No no, I don't think you'd need it'—in response to question about awareness assist; intermediate group, participant 1, line 510). It may be that learners, with less skill, have a greater need for awareness assist, which is supported by research showing a decreased ability to detect hazards (Mayhew & Simpson, 1995; Quimby et al., 1986). Advanced drivers made no changes to their interface design; however, as their design for the 1-h scenario was relatively brief, this is unsurprising given their previously stated views on awareness assist.

7.4.3 Comparison of Groups' Handback Designs

Each group requested a way of visualizing the state of automation and its capabilities. Learner drivers suggested that the car presented its proposed future intentions, intermediates suggested a HUD that showed that the car could detect other vehicles, and the advanced group favored a simpler design consisting of a traffic light system on how risky it is to activate automation given current infrastructure and road condition. This kind of information could enable drivers to calibrate trust with automation so that they can assess whether they can rely on automation prior to activation, a concept well reported across research into how operators use automation (Lee & See, 2004; Walker et al., 2016). Both the learner and advanced group requested a mode change either on the HDD or HUD for clear awareness that the automation is in control, which addresses the problem of mode error that may come about through interactions with automation (Sarter & Woods, 1992, 1995; Stanton et al., 2011).

Intermediate drivers requested more strategic information prior to handback such as a confirmation of destination and junction, a display to show how long the system can be automated for, and other cues such as road signals to show that automation can be activated.

7.4.4 RELEVANCE TO DSA AND JA

Although participants were not directly asked to design around DSA and joint activity (JA), discussing these findings in relation to these theories will provide value for the remaining chapters. By revisiting Salmon et al.'s recommendations in light of DSA (Salmon et al., 2009), a number of comments reflected a need to tailor interactions to drivers. Quotes such as 'you don't want it to say too much to you, you know like a voicemail message, it takes forever to get the information you need' (intermediate group, participant 5, line 149–150) indicate a need for information to be well tailored and efficiently delivered to the driver during transactions. Another example of addressing the DSA recommendations is the use of customizable interactions—a recurring theme throughout all focus groups.

Differences across groups provide an indication that driver schemata vary across skill groups and each require a tailored approach to C/HAV handover and handback. Examples include the expression of favor toward HUDs and HDDs during handover assist in both learner and intermediate drivers when compared to advanced drivers. As illustrated in previous research, advanced drivers may possess schemata that allow them to raise their own SA more readily than that of their less experienced counterparts (Bainbridge, 1978; Walker et al., 2009).

As a final example of an insight of applications of DSA in C/HAV interactions, many drivers across skill groups requested a handback display that presents the driver with the performance, awareness, and the future plans of the automated system to be present prior to the driver relinquishing control. Access to information such as this, at this particular time, indicates how both driver and automation can possess different but compatible SA, which is then communicated at a specific time point to facilitate decision-making of drivers—i.e., whether to activate automation (Sorensen & Stanton, 2016; Stanton et al., 2007).

These focus groups indicate that DSA, when applied to C/HAV handover/handback, seems to address a number of drivers' recommendations for design—this may be in part due to its ability to address safety, efficiency, and tailored designs toward individual requirements and the level of skill (Salmon et al., 2009).

From the perspective of JA, aspects of the handover interactions provided in this chapter relate to JA's core concepts (Klein et al., 2004, 2005). For example, much of the focus of the handover interactions provided is on 'what am I expected to do?'—related to timings and actions in the future. This addresses the JA concept of directability, an important aspect to ensuring that human and machine teams work effectively together. Examples include ensuring that alerts are given in a timely manner and direct the driver to the correct inputs when expecting to transfer control.

During handback, all groups provided a recommendation related to the capacity of the vehicle and whether it was safe to activate automation under current circumstances. For example, the learner group requested communication related to what the vehicle is going to do within the next 1 min and to indicate when it is fully safe to transfer control. The intermediate group requested a visual indication of what the vehicle can see, so that they can ensure that the vehicle is detecting both vehicle and weather appropriately before transferring control. Finally, the advanced group requested a traffic light system, indicating when activating automation was either risky or completely safe. Each of these suggestions address the vehicle's ability to perform its given role, and by communicating such information, the driver can make better decisions regarding safe operation of the vehicle.

Confirmations and on-the-fly changes to route plans indicate the requirement to reach shared goals and expectancies while the HUD and HDD serve as coordination devices, aligning actions in real time. Together, these design recommendations address the concepts of JA. Going forward, ensuring that these specific recommendations are prioritized in design is important, as these recommendations are likely to be effective in coordinating activity between driver and automation.

7.4.5 LIMITATIONS

As a focus group study, the designs presented represent the approaches taken between groups, which may vary between individual groups regardless of driver experience category. As an example, the learner group preferred to list out multiple ideas and then debate them, whereas the intermediate group agreed on each element prior to moving on to the next stage. Further, the amount of disagreement varied from group to group, with advanced drivers showing the most disagreement regarding individual solutions, and this led to variation in the groups' abilities to converge on an agreed solution. In response to disagreements, the groups decided on methods of customization.

A further limitation to note is that the scenarios found within this study are focused on UK roadways, and participants recruited for the study were living within the UK for study or work. It follows that the designs, responses, and recommendations made by participants may be more aligned to environments found within the UK (e.g., weather conditions, road layout, distance between junctions, hazards, national law, and the behavior of other road users). Further work will be required to gain a better understanding of what different users from various nations across the world require from AV interaction.

Finally, each skill group only featured one focus group worth of data. The sample, therefore, may limit potential applicability of these findings as they appear in this study. However, these recommendations can still serve as a guide and discussion point for further developments, particularly toward how different individuals exhibit different needs during C/HAV interaction.

7.4.6 CONCLUSION

This study explored how drivers of differing skill levels approached the handover problem through the use of handover HMIs. HMI solutions for the automation-to-driver transition of control from three focus groups (learner, intermediate, and advanced drivers) were presented. Additionally, amendments for shorter TOOTL and three solutions for the driver-to-automation transition of control were outlined. As predicted, advanced drivers show a preference for limited information in awareness-assist interfaces and generally preferred not to rely on HDDs and HUDs for transitions, whereas learner and intermediate drivers requested more information to guide them through the handover using multimodal approaches. It is worth noting that advanced drivers may be exhibiting more complacency when recommending lower levels of driver assistance (Charlton & Starkey, 2011). For this reason, advanced drivers may in part be more capable of performing but may not rely on assistance systems in the process. Therefore, care must be taken to ensure that 'need-to-know' information is delivered to all drivers of varying skill, leaving nonessential information to be customizable.

Customization of handover protocol was a common theme throughout the discussions involving changes to alert times and changes to modalities for the display of information. Innovative designs for handover and handback were created including surround cameras and augmented traffic situation (learner group), sophisticated timing systems (intermediate group), and ghosting control of the vehicle prior to handover (advanced group). When addressing the handback, factors such as calibrating trust (Lee & See, 2004; Walker et al., 2016) could be assisted by providing the driver detailed information about automation performance and intentions.

This study provides researchers and designers with ideas for the transition of control in noncritical scenarios, as well as an appreciation of how this may differ with greater driving experience. Implementing designs to accommodate for new insights into DSA shows that tailoring the handover toward individual requirements in the driving task is of great importance (Salmon et al., 2009). Further, as drivers of differing skill vary based on attentional resources (Young & Stanton, 2007a) and capability to develop and maintain SA (Walker et al., 2009), it is important to address this factor when considering handover HMI designs. Further work should explore the role of customization for application to C/HAV design—however, care must be paid as to what information remains safety-critical information during transitions of control and responsibility.

7.4.7 FUTURE DIRECTIONS

This chapter presented issues in C/HAV interaction to users of varying skill level. The next chapter develops designs by factoring in all previous chapters, including the results from the participatory design workshops.

8 Designing Automated Vehicle Interactions Using Design with Intent

8.1 INTRODUCTION

To give structure to the design process, this chapter culminates all previous chapters by presenting previous findings to a group of human factors experts and conducting a workshop centered around a human factors group design technique—design with intent (DwI; Lockton & Stanton, 2010; Lockton et al., 2010). Many methods are available to practitioners to analyze the domain under analysis and generate design guidelines that are both in line with theory and address practical concerns (Stanton et al., 2017a); however, methods that are broad and efficiently generate design solutions are few and underrepresented in the literature. The DwI toolkit (Lockton & Stanton, 2010; Lockton et al., 2010; Lockton, 2015), a concept generation tool, is among these underrepresented methods. Its inception in 2010 saw great traction in human factors research (376 Google Scholar citations at time of writing). Although influential, its application is rarely reported; consequently potential insights into the power and scope of the tool remains largely unseen.

8.1.1 INTRODUCTION TO DESIGN WITH INTENT

The DwI philosophy postulates that desired behaviors can be encouraged through design and that any given system design can benefit from interdisciplinary design methods (i.e., not necessarily domain specific methods), which is largely beneficial to new domains with little accumulated knowledge (Lockton & Stanton, 2010; Lockton et al., 2010). The DwI toolkit compiles 101 design considerations for guiding human behavior, broken down into eight lenses—each addressing a different 'worldview' expressed across a variety of domains. Each card presents the individual with a consideration such as from the cognitive lens—'Habits—Can you make it easy for a new behavior to become habitual, by building it in to an existing routine?'. Individuals record their responses, with the given problem and solution in mind. For example, a car manufacturer may wish to place their headlight switches closer to the ignition to encourage automaticity when switching off lights before leaving the vehicle (see Figure 8.1 for example).

Without prompts like these, designers may overlook opportunities for safety-critical behaviors to become managed and guided. Although typically conducted as a group, there is no strict way of applying the toolkit. The authors indicate that the toolkit is not a prescriptive method and serves to guide discussion rather than dictate

DOI: 10.1201/9781003213963-11

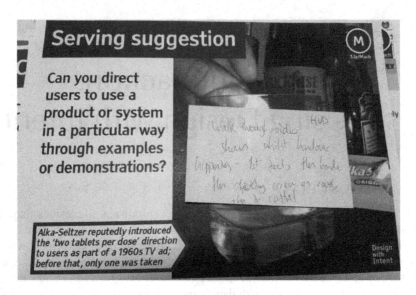

FIGURE 8.1   Example card found in the DwI toolkit.

it. It follows that designers from a variety of domains can adapt the toolkit around their specific requirements and roles.

8.1.2   DESIGN WITH INTENT FOR IN-VEHICLE INTERFACE DESIGN

An application of DwI for improving in-vehicle display interfaces comes from Allison and Stanton (2020). Allison and Stanton (2020) utilized the DwI toolkit to facilitate creativity in relation to displays that communicate fuel use, cost, and emissions to reduce driver carbon footprint. Categorized into human–machine interface (HMI) modality and theme, the authors present 138 design suggestions related to human interaction and the promotion of fuel efficiency. Through a process of exclusion, design suggestions were reduced to a total of 14. These were then rated by participants to gauge user likelihood to use the design specification. Allison and Stanton (2020) demonstrate the effectiveness of DwI by clearly and succinctly summarizing workshop findings their application using quantitative and qualitative methods.

The task of optimizing interactions between driver and automation in level 3 and 4 automated vehicles (AVs) can be achieved through considering interface design, alongside temporal aspects of the AV domain such as pre-journey, manual driving, transfers of control and automated driving. By focusing on the design goal—an automated driving assistant that collaborates with the driver—issues such as mode errors and coordination issues could be better prevented as well as improving situation awareness (SA) (e.g., Ackerman et al., 2017; Flemisch et al., 2012; Klein et al., 2004, 2005).

This is thought to be achieved through concepts such as being able to interpredict one another's actions, be aware of system capacity, and direct one another toward actions and areas of interest. This book proposes that this should occur throughout the automated cycle to better align goals and expectancies so that system performance is improved throughout the journey. To generate concepts related to these values, DwI can aid in putting these values to practice by suggesting novel approaches toward communication, either through explicit displays or through wider means such as vehicle decision-making, culture, and education.

8.1.3 Current Application

This DwI workshop brought together a group of human factors experts to consider the problem of human–automation communication in level 3 and 4 automation. The goal of the workshop was to consider each of the 101 cards and discuss potential solutions or ideas related to the card to design an automated assistant that guides transitions of control, relays capacity, situation information, and aims to improve clarity, usability, and optimize trust. Participants were given an overview of the preceding chapters to help guide discussion and center recommendations around preexisting findings within this body of work. The handover of control was framed as a major concern of level 3 and 4 AVs, but the assistant was to be considered throughout the automation cycle. We present themes and a design solution generated from this workshop to illustrate the following: 1) When considering the 101 prompts from the DwI toolkit, what are deemed to be the most important considerations for human factors experts toward shared-control AVs. 2) To provide an example method for utilizing the DwI toolkit for prototype inspiration within a safety-critical domain.

The process for conducting the workshop featured a divergent stage and a convergent stage. During the divergent stage, participants went through each card and generated design suggestions in line with how the card could address the findings within the book's previous chapters. Once complete, the convergent stage drew together recurring and similar suggestions into a single example design to demonstrate how a coordinative automation assistant could manifest itself physically (see Figure 8.2 for an illustration of the process).

8.2 METHOD

8.2.1 Participants

Ethical approval was granted via the University of Southampton ethics panel (ERGO II No. 47643). Five participants (2F, 3M) were recruited through internal communications at the university. All five participants were experts in human factors with experience of working in automobile domains (two of which were AV researchers). Participants had been researchers in human factors for a mean of 12.4 years (range of 4–34 years).

8.2.2 Design

The DwI workshop lasted a full day, with 5 h allocated to generate design suggestions for each of the 101 workshop cards. This involved each participant writing

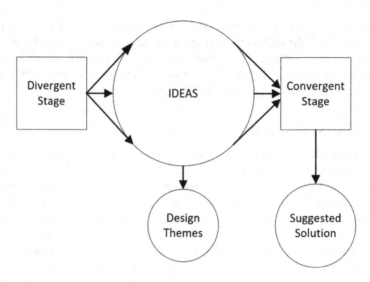

FIGURE 8.2 DwI workshop outline—divergent and convergent stages and their outcomes.

comments for one fifth of the cards dealt out (approximately 20 cards) and feeding back their ideas to the rest of the group. When presenting back to the group, other members were able to provide additional ideas following discussions. These comments formed the basis for thematic analysis.

8.2.3 MATERIALS

A presentation on level 3 and 4 AVs and the current state of the domain was delivered to the workshop. The presentation focused on the theoretical frameworks outlined in Chapter 2, the current state of AVs, and the work completed previously. The 101 DwI cards were used for idea generation and discussion. To aid in discussion surrounding the optimization of communication, the abstraction hierarchy presented in Chapter 3 from the cognitive work analysis (CWA) (Rasmussen et al., 1990; Vicente, 1999) provided the group with an idea of what values effective communication aimed to address and the physical objects that are available to designers in this domain (e.g., head-up displays, vibrating seats, and lights on steering wheel). The SOCA-CAT allowed the experts to see what tasks were required to be performed throughout the automation cycle and which agent was responsible (e.g., navigation and who is in control and when). An overview of previous work and findings was also given, alongside copies of figures from previous studies to align participants with the objectives and previous work outlined within the book. How interactions took place and what was to be communicated during interaction with the system were completely open for participants to decide.

8.2.4 PROCEDURE

Due to the niche nature of the target sample, participants were recruited internally and opportunistically. Participants arrived at 9 am on the day of the workshop. They were welcomed, briefed, asked to read the information sheet, and sign the consent form to indicate that they are content with the day's activities, anonymity, confidentiality, and data protection. Following this, a safety brief was conducted as well as a brief introduction to the researcher and an introduction to the field of research was delivered by the lead researcher.

The presentation lasted for 30 min and involved several aspects of automated driving such as the levels of automation (specifically level 3 and 4 automation, in which this research considers), the drawbacks of these levels, an introduction to joint activity, and distributed situation awareness (DSA) in which the tenets outlined in Sections 2.1.4.1 and 2.1.4.2 were described in detail, a broad overview of previous notable findings, the CWA, the findings from previous chapters, and what yet needs to be developed. The presentation was supplemented with printouts of notable figures throughout this book and a list of tenets from both JA and DSA. These printouts remained present during all stages of the design workshop to allow participants to refer to them during the task.

The work domain analysis and the SOCA-CAT was central to the discussions in this workshop. These along with figures from previous chapters outlining experimental findings were given to the group prior to the session and were guided through the content during the presentation. They were then introduced to their design problem: 'to consider how to optimize interaction between driver and automation where transfers of control are to be expected'. To aid in this task, the outcome of a previously constructed CWA output shows why effective communication is important—notably, to improve usability (Barón & Green, 2018; Nwiabu & Adeyanju, 2012; Ponsa et al., 2009; Schieben et al., 2011), SA (Stanton & Young, 2000; Heikoop et al., 2016), efficiency, trust (Lee & See, 2004; Walker et al., 2016), coordinated activity (Bradshaw et al., 2009; Klein et al., 2004, 2005), safety (Brandenburg & Skottke, 2014; Merat & Jamson, 2009; Eriksson & Stanton, 2017b), and optimize workload (de Winter et al., 2014; Young & Stanton, 2002b). Using the CWA findings from Chapter 3, participants were introduced to what is available to HMI designers for the purpose of facilitating interaction, such as center consoles, head-up displays, and vocal communication.

With an understanding of the field and the questions that are required to be addressed, participants were then individually dealt out a selection of cards (randomized) from the DwI pack (5 participants—approximately 20 cards each). Individually, participants wrote on post-its notes on their cards in a way in which they believe the card could contribute to an optimal interaction between driver and automation—keeping the design values in mind throughout. Upon completing their individual cards, participants presented their card to the group along with their thoughts surrounding their suggestions. Other members of the group then added additional post-it notes to these cards as discussions unfolded. Following this, the participants were thanked for their time and then debriefed.

8.2.5 METHOD OF ANALYSIS

The data from this workshop comes directly from the comments related to each of the 101 cards. These comments were categorized using thematic analysis. The thematic analysis considered the topic that the comment referenced. Where communication was concerned, this was subcategorized to allow for a thorough exploration of the issues that were deemed important during the workshop.

8.3 RESULTS

This section provides an insight into the theme categories for level 3 and 4 AV interaction that HF experts commented on, along with example comments from the workshop.

8.3.1 THEMES GENERATED DURING DIVERGENT STAGE

Table 8.1 and Figure 8.3 shows the themes and distribution across comments. The X-axis indicates the theme denoted to each comment, and the Y-axis indicates the frequency at which comments were coded under each theme. Communication was the most discussed element regarding interaction design. Next were implementing boundaries toward how the AV should operate and comments related to the social aspect of automated driving (e.g., interacting with the community and connected vehicles). Finally, four themes were rarely referenced, including learning materials for AV operation, how the vehicle should behave under certain conditions, how the physical layout of the vehicle should operate, and how the interaction should be customizable. Each theme is discussed in detail in the following sections.

8.3.1.1 Communication

Communication made up nearly half of comments made during the session (50.7%). The theme communication relates to any comment reflecting the nature of the HMI, how the vehicle should communicate with the driver, and how SA should be raised during transitions.

Due to the breadth and complexity of this theme, this theme was broken down into six subthemes to better navigate the comments made. Table 8.2 and Figure 8.4 outline these subthemes by outlining their frequencies, descriptions, and example quotes to illustrate each subtheme.

8.3.1.2 Boundaries

Returning to Figure 8.3, the second most discussed theme was the use of boundaries (14% of comments). This involves the strict implementation of rules or the blocking of control functions under certain conditions. These conditions may include driver unreadiness, situational constraints such as terrain, or amount of driver experience. This is illustrated by the following quote: 'Manual driving only operates when driver is in correct position; Errorproofing—Matched Affordances'. These comments indicate a condition that may be breached and the boundaries that the system implements in these situations. These conditions may relate to driver condition but also

TABLE 8.1

Frequency of Suggestions within the Themes Generated and Their Respective Lenses

	Architectural	Cognitive	Errorproofing	Interaction	Ludic	Machiavellian	Perceptual	Security
Behavior	1	0	1	0	0	3	2	1
Communication	13	10	5	12	5	5	14	8
Customization	1	1	2	1	1	0	0	0
Implement boundary	7	1	7	0	0	4	1	4
Learning	1	3	0	1	2	2	0	1
Physical	4	2	1	0	0	1	0	0
Social	0	4	0	2	5	3	2	3

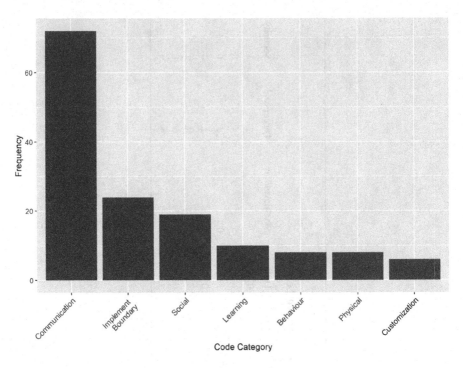

FIGURE 8.3 Bar chart showing themes generated from thematically analyzing DwI comments and their associated frequency of comments.

automation capability, as illustrated by the following quote: 'Automation only available in places that work; Security—Where you are'.

8.3.1.3 Social

Linking the driver to the social domain was deemed important by HF experts (11% of comments). Comments relating to the 'social' theme encompassed legal recommendations (e.g., 'Driver should always have ultimate responsibility; Perceptual—Watermarking') or connected the driver to the social environment (e.g., 'Customization sharing—for different road conditions, e.g., genius play tests—different events different environments; Ludic—Make it a meme'). The social theme suggests that automated driving behaviors can be encouraged using peer-feedback or social coercion.

8.3.1.4 Learning

Education and learning how a system operates feature in the 'learning' theme (5.8% of comments). Experts commented on how skill will develop over time with continued use, and with it, conditions can change. This may be through bottom-up approaches to learning (e.g., 'Make automation availability consistent—road condition, road type, traffic. So, users learn pairings; Cognitive—Habit') or through top-down methods such as training programs or introductory materials (e.g., 'Video's/

TABLE 8.2
Subthemes for the Theme 'Communication' Along with Descriptions and Example Comments for Each Subtheme

Subtheme	Description	Example comment
Coordination	Collaboration of timings, states, and guidance toward what is expected during the automated cycle	'Checklist or sequence "have control", "take control", "control accepted", "OK". Protocol for info transfer like aviation'. Perceptual lens—implied sequence
Awareness	The raising of situation awareness within the system. Either through HMI displays or communication that relates to the environment (e.g., hazards)	'Radar based damage trajectory for maneuvers'. Security lens—threat to property
Coaxing	The concept of invoking a preconceived behavior for the driver to perform. For the most part, this relates to engaging with the system rather than ignoring it	'Driving manual in automation zone gets lots of requests for automation'. Machiavellian—poison pill
Connectivity	The personification of an automated driving assistant or improving the ease of accessibility with AV interaction	'Personas for the automated system; Cognitive—Emotional Engagement' or 'Level 5—taxi—Level 4—guardian angel—Level 3—driving instructor—Level 2 backseat driver with human controls—extension of persona idea'. Perceptual—metaphors
Alerting	Direct commands such as notifications and alerts—typically to avert potential unawareness of the driver	'Warn if driver inattention is detected—automation watching over driver when they are driving manually and warns them if error is detected'. Errorproofing—conditional warnings
Capacity	The automated vehicle's or driver's ability to handle a current or upcoming situation. This may involve relaying performance metrics as automation behaves or notifications of potential uncertainty in whether the agent can handle the task it is expected to perform	'Emphasize limits or capability of automation -> what it can or can't do -> educate drivers on limits'. Cognitive—scarcity or 'Traffic light indication of increasing readiness of automation to take control'. Interaction—partial completion

animation of polite role model users. Accessed via center console operational at the setup stage; Ludic—Storytelling').

8.3.1.5 Behavior

How a vehicle or driver behaves at a certain time was another aspect of interaction according to HF experts (5.2% of comments). Here, there were no specific messages relayed between either party; merely, they would act in a certain way under certain conditions. Comments are typically related to vehicle behavior (e.g., 'Form of

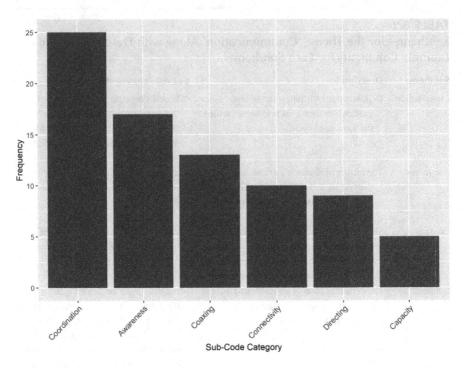

FIGURE 8.4 Bar chart showing themes generated from thematically analyzing DwI comments and their associated frequency of comments.

steering wheel changes based on automation readiness/control and human sensors—'normal' steering wheel only when completely ready for driver control'; perceptual—perceived affordances), or decision-making (e.g., 'car drops itself down levels of automation (however, could cause mode confusion)'; architectural—segmentation and spacing).

8.3.1.6 Physical

Physical properties, particularly prior to handover, make up the 'physical' theme (4.7% of comments). Physical properties involved control elements such as button inputs (e.g., 'Button to give control to human based on steering wheel—2 buttons'; architectural—positioning). Other comments are related to the body positioning and seat configuration when a handover is to be expected (e.g., 'Manual driving only operates when driver is in correct position'; errorproofing—matched affordances).

8.3.1.7 Customization

The final theme for comments during the divergence stage of the DwI workshop involved the presence of customizable functions in AV interaction (3.5% of comments). Comments related either to automatic setting changes decided by the AV (e.g., 'auto customize options/functions for different contexts and environments';

errorproofing—choice editing) or through driver-initiated customization (e.g., 'Amount of control, e.g., lane keeping—little or full support—implement dials'; errorproofing—portions).

8.3.2 Solution Generated for Convergent Stage

Following the divergent stage, in which comments were provided for each of the 101 cards in the DwI toolkit, experts took part in the convergent stage, where concepts home in on a single response to the factors that were considered prior. The group created an automated assistant named 'Steeri', which coordinated with the driver depending on context and stage of the automation cycle (see Figure 8.5). In the event of planned or unplanned handovers, Steeri provides the driver with real-time assistance to navigate plans, alerts, and monitor driver awareness. The context of automated driving and the capacity that automation has was an important aspect of this design. Steeri communicates with the driver through an augmented head-up display and vocal interaction. Steeri raises the awareness of the driver through real-time hazard identification and notifications for transitions of control. Depending on current operational capacity, Steeri would change its features (through the medium of fashion, i.e., a captain's hat when in control). This design makes use of many levels of automation, suggesting that vehicles could feature level 1, 2, 3, and 4 capabilities. Here, the driver could have one feature automated (e.g., lane keeping) whereas, if all functions are automated, Steeri can communicate to the driver when he needs to be monitored in case of a potential dropout. State would be primarily communicated through a state bar with the vehicle indicating current automation state. Steeri features trip planning and navigation capabilities, with the driver having the option of requesting Steeri to 'take me to work'. A summary of the components found within the convergent solution is presented in full in Table 8.3, along with the related themes from the analysis.

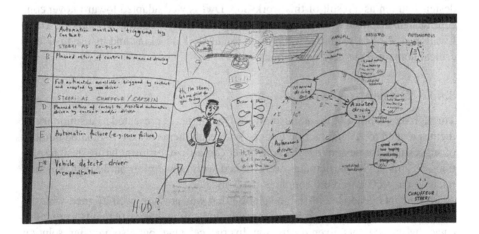

FIGURE 8.5   Convergent solution following the divergent stage.

TABLE 8.3

Features of Divergent Solution, Description, and Associated Themes

Design Component	Description	Associated Theme/Subtheme
Adapts to performance capacity	Automation transitions between modes dependent on situational constraints	Boundaries, behavior, capacity, coaxing
Two-way dialogue	Vocal cues and alerts coordinate actions and allow the driver to interact with Steeri to raise situation awareness	Alerting, coordination, awareness, connectivity
Anthropomorphic assistant	Driving state and situation is communicated through an assistant that represents a copilot	Connectivity
Icon indications for mode	Steeri changes image based on mode and situation	Alerting, awareness, coordination, learning
Augmented situation information on HUD	Driver can access situation information close to the road environment	Awareness
Bar indicating state	A sliding bar allows the driver to see where on the scale the current driving setting is	Awareness, coordination, capacity
Psychophysical sensors	Sensors can detect whether the driver is able to drive	Awareness, capacity, coaxing

8.4 DISCUSSION

The DwI toolkit serves as an idea generation tool for product or system designers to consider the complexity of human interaction (Lockton & Stanton, 2010; Lockton et al., 2010). This work demonstrates that although DwI originated from product design, it can be readily adopted to deal with new safety-critical domains where guidance is limited. By utilizing 101 cross-domain considerations this study provides a demonstration of how DwI can be used to identify potential implementations of technology in the target domain of level 3 and 4 AVs and provides an example design solution as a result of this workshop. DwI is well adapted toward novel concept design inspiration in newly developing domains (Lockton & Stanton, 2010; Lockton et al., 2010); however, published use of DwI is lacking and its use in safety-critical domains is largely undocumented.

8.4.1 OVERVIEW OF FINDINGS

Through the medium of thematic analysis, DwI output is presented as a summary of themes related to AV interaction design. With each theme are an associated number of suggestions for design that the researcher can utilize for formulating design solutions. In this study, participants were tasked with converging on a single solution related to aspects covered throughout the workshop.

Participants generated a design solution that was not only well aligned with the themes identified in the workshop comments (indicating that participants were influenced by the prompts given during the divergence stage) but also a design solution that merges together a vast number of previously researched concerns in level 3 and 4 AVs. Experts suggested the use of a personified virtual assistant which can communicate emotional cues and engage with the user in dialogue. This, in tandem with coordinative communication (i.e., state displays and guidance through transitions of control), awareness communication, alerts, and careful planning, goes a long way toward addressing the subthemes identified within the theme 'communication'. The automated assistant, due to its personification, could connect the driver with automation by communicating feelings such as happiness, worry, or inactivity in certain situations—in line with previous work in AV interaction. Not only this, but an automated system could increase safety by determining where and when automation can function and whether to handover control to a driver in certain states of unawareness—addressing the theme 'boundaries'. If this system were able to guide and teach the driver and eventually allow them to adapt settings as skill improved, then the solution would address two additional themes—'learning' and 'customization'. The system could monitor driver readiness in terms of physical position and include features that reduces mode error or physical errors such as slips—addressing the 'physical' theme. This design goes some way toward addressing the issues of increased automation capabilities such as mode confusion, degraded SA, and agent coordination (de Winter et al., 2014; Endsley & Kiris, 1995; Eriksson & Stanton, 2017b; Sarter & Woods, 1992, 1995; Sorensen & Stanton, 2016; Stanton et al., 2017b).

This design solution is just one example of taking the first steps into designing a user interaction in a safety-critical domain. It is apparent that many of the features found within the convergent solution relate to the comments made during the stage prior. Going forward, researchers with output like this may wish to create a prototype that resembles the convergent solution(s) from the design workshop and test them against current designs.

8.4.2 Applications to Future AVs

Understandably, in this study, human factors experts identified solutions focused on communication concepts such as HMI displays to address AV interaction. The subthemes identified for the theme 'communication' fall in line with previous research, notably where interfaces explicitly communicate state, performance, and capacity information to improve AV interaction (Beller et al., 2013; Klein et al., 2004, 2005; Lee & See, 2004). A similar concern was directed toward with the states of awareness within the system and make an array of suggestions toward raising SA though transactions of information (Ackerman et al., 2017; Endsley & Kiris, 1995; Sorensen & Stanton, 2016; Stanton et al., 2017b). Further, the role of function allocation (Fuld, 2000; Idris et al., 2016) in this study was largely centered around the capability of an AV to work in a given environment or lockout a driver who is not deemed aware or capable of taking control of the driving task. At the time of writing, current AV designs from manufacturers do not integrate communication features such as these in great detail (Audi, 2019a; Cadillac, 2020; Tesla, 2018, 2020).

These findings show that is becoming increasingly evident that to address the various issues with shared-control AVs, automated assistant must strike a balance between addressing all concerns and not overloading the individual with redundant information that either masks important information or overloads mental capacity

(Brickman et al., 2000). DwI goes some way to alleviate these issues, as considerations are made concurrently with a view on a converging design solution.

8.4.3 Relevance to DSA and JA

Experts within this workshop were guided by the chapters outlined previously in this book. They were introduced to DSA, JA, and the CWA presented in Chapter 3 as part of their orientation presentation (inspired by Stanton et al., 2006, 2017b; Salmon et al., 2009; Klein et al., 2004, 2005). Further, these materials were made available to experts during the divergence and convergence stages. In particular, the theoretical frameworks of both DSA and JA, along with the CWA from Chapter 3, provided experts with foundation upon which designs could be based upon. This information, along with the experience that experts brought to the design process, was structured in a way so that automation designs could optimize communication throughout the automation cycle.

The design outcome generated by the workshop appears to be well aligned with these theoretical frameworks. Steeri represents an automation assistant that can relay messages in real time and plan for activities in the future. With such capabilities present in AVs, communication can progress flexibly and at the pace of each agent using vocal communication as the primary method of interaction. Table 8.4 associates Steeri's design components with the associated principles of both DSA and JA (Stanton et al., 2006, 2017b; Klein et al., 2004, 2005).

Table 8.4 serves as a discussion point regarding the design output. Steeri has multiple states that allows the AV to perform at different levels by taking on varying degrees of automation regarding current capacity. This ensures that the AV is operating safely

---

TABLE 8.4
Design Components and Associated Theoretical Principles Outlined in Sections 2.1.4.1 and 2.1.4.2

Design Component	Associated DSA Principle	Associated JA Principle
Adapts to performance capacity	—	Capacity
Two-way dialogue	Supports transactions, customizable, explicit communication link, facilitates SA	—
Anthropomorphic assistant	Supports transactions, explicit communication link, facilitates SA	Common ground, capacity
Icon indications for mode	Facilitates SA	Phases, capacity, coordination devices, mutual predictability
Augmented situation information on HUD	Supports transactions, explicit communication link, facilitates SA	Mutual predictability
Bar indicating state	Supports transactions, explicit communication link, facilitates SA	Phases, capacity, coordination devices, mutual predictability
Psychophysical sensors	—	Capacity, mutual predictability

Source: Derived from Salmon et al. (2009) and Klein et al. (2004, 2005)

and is communicated to the driver via a head-up display. Next, by allowing Steeri to engage in a two-way dialogue, transactions are supported by both human and automation having the ability to access information related to their own mental models. In particular, this two-way communication supports SA by allowing both agents to adapt to the current context of the road environment. An anthropomorphic element for Steeri coupled with the two-way vocal interaction outlined within the design will allow drivers to seamlessly build common ground while being able to access information regarding capacity in a more direct and understandable way. Anthropomorphism has also been found to be beneficial to calibrating trust in AVs indicating that this approach may be beneficial (Waytz et al., 2014), a feature that Steeri possesses.

The visual displays featured within this design allows the driver to access information regarding how well the vehicle is performing under certain constraints. By being able to monitor the situation in this way, the driver is better able to predict the future behavior of the vehicles and adapt their actions accordingly. This approach therefore facilitates SA and provides an explicit communication link due to the augmented display residing within the driver's environmental view. A state bar would allow drivers to quickly gain information regarding the capacity and status of the vehicle which would in turn allow for better system SA. Finally, physiological sensors would allow for the vehicle to directly address the capacity of the driver prior to transitions of control, therefore allowing the vehicle to intervene appropriately.

Going forward, this design appears to address the core concepts of both DSA and JA and would therefore be appropriate for testing. This design, however, does not address the handover information transfer and physical collaboration in a detailed manner but rather focuses on global approaches to driver–automation interaction. Therefore, previous work from user design workshops and experimental testing will be integrated with this design within Chapter 9 to create a complete prototype that can be tested in comparison with current AV HMI.

8.4.4 DwI and the Future

This study shows the benefits that the DwI toolkit has for the developing safety-critical domain of level 3 and 4 AVs. Individually, themes generated in this chapter relate to issues that have faced automation since its inception. However, due to the nature of controlled experiments, research into level 3 and 4 AV interface design typically addresses issues in isolation with little insight into how features should be integrated. In this study we provide examples of rich qualitative data and a design solution from the use of DwI that can aid in the prototyping process. DwI's versatility may reside in its ability to pull together multiple considerations across multiple domains from many years of research. Not only does this mean the tool is flexible to the target domain but also detailed in its scope. Going forward, it is important to document the methods in which DwI is applied to certain domains.

8.4.5 Conclusion

DwI is a versatile method for generating design solutions for the purpose of addressing human interaction. Its creators state that it can be used across domains, in a way that researchers feel to be appropriate. This flexibility makes it ideal for kickstarting the prototype stage of a newly developing domain. We provide in-depth results from a DwI workshop involving five human factors experts, with the design problem of creating an automated assistant in shared-control AVs. Our analysis identifies a wide range of themes that were deemed to be important during the workshop and presents a solution as a result of the 101 considerations made during the workshop. The final design from this workshop collates factors such as communicating state, awareness, capacity, and performance, as well as guiding the driver through transitions, upcoming events while making sure to connect with the driver through personified interfaces. Our findings show that in a short period of time (one workshop session lasting 7 h) experts can identify high priority themes, provide suggestions, and converge on a design solution that can be adapted for further testing and demonstrate that future designers in high-risk domains should consider the role of DwI in opening discussion and provide practitioners with a bridge between domain analyses and practical solutions.

8.4.6 FUTURE DIRECTIONS

Chapter 8 provides design recommendations and an example handover assistant to address the appropriate outcomes of a collaborative handover assistant. All previous chapters contribute to the design process. Chapter 9 finalizes the book by presenting a handover assistant that falls in line with previous findings—making use of two-way interaction, vocal and visual communication, allocation of tasks, and ensuring that communication is collaborative and occurs throughout the automation cycle. Chapter 9 tests prototypes to validate the design using rigorous testing methods and presents findings on how the novel handover assistant improves interaction between driver and automation.

Section IV

Testing and Validating a
Novel Prototype

9 Validation and Testing of Final Interaction Design Concepts for Automated Vehicles

9.1 INTRODUCTION

To bring together the findings from this book and show how the design recommendations from previous chapters contribute toward positive outcomes in human–automation interaction, this chapter presents the development of an automation assistant prototype and presents findings from a validation study conducted in a driving simulator. The prototype presented in this chapter represents the culmination of all preceding chapters. The following sections provide a summary of research in handover assistants, a summary of the design process found within this book thus far, a description of the proposed prototype, and the technical development of the automated assistant prototype.

9.1.1 SUMMARY OF AUTOMATION ASSISTANTS

In Chapter 2, an introduction to conditionally and highly automated vehicle (C/HAV) assistants outlining the key endeavors of the C/HAV research community over recent years was provided. This section provides a recap of key literature and discusses the main contributions for improving human–automation interaction from this book.

As stated in Chapter 2, areas that have been explored in C/HAV interaction include situation awareness (SA) (Merat & Jamson, 2009; Stanton et al., 1997), event notifications (Bazilinskyy & de Winter, 2015), time to takeover (Eriksson & Stanton, 2017a; Gold et al., 2017; Young & Stanton, 2007b; Zeeb et al., 2015), effect of demographics on performance (Körber et al., 2016), effect of traffic density (Gold et al., 2016), effect on driver behavior (Merat et al., 2014; Naujoks & Neukum, 2014; Naujoks et al., 2014), and distractions (Mok et al., 2015). Direct interface solutions have explored the following: 1) alerts informing of situation and takeover time (Walch et al., 2015), 2) exploring multimodal alerts and the effect of direction on takeover performance (Petermeijer et al., 2017a, 2017b; Walch et al., 2017), 3) ambient and contextual cues to facilitate takeover (Borojeni et al., 2016), 4) graded takeover request in 'soft takeover request' scenarios (Forster et al., 2016), and 5) communicating urgency information (Politis et al., 2015). Although many of these behavioral measures have for the most part been confirmed and accepted among

DOI: 10.1201/9781003213963-13

automated vehicle (AV) HMI designers, there is currently a gap in how information regarding the situation and environment should be communicated to the driver prior to them regaining control. Further, providing a solution that draws together multiple considerations that directly addresses system (SA) and many other target outcomes is yet to be provided.

The research outlined here explore singular outcomes and design issues that contribute to safe C/HAV operation. However, the majority of previous research does not consider these individual elements as part of an overall automation interaction throughout a journey in a real-world driving environment.

As outlined throughout the book, positive outcomes for C/HAV interactions can be summarized with the following requirements for a C/HAV digital assistant:

- Regaining SA lost due to the driver being 'out of the loop' during automation.
- Identifying and communicating safety-critical information related to the event and the actions that are required to be performed.
- Communicating mode of driving and capability to operate safely.
- Guiding the driver toward physical objects required to safely operate vehicle.
- Ensuring optimal usability for the driver.
- Calibrating trust to prevent misuse and disuse.

To address this multifaceted design issue, research into the safe implementation of automation suggests that automated agents are required to adhere to cooperative principles—summarized by Klein et al.'s (2004, 2005) articles, extending Clark's (1996) pivotal contributions regarding the use of language and grounding for effective cooperation. Klein et al. (2004, 2005) outline the requirement for interfaces to communicate intentions, capacity to perform, phases, and coordinate actions. In doing so, they posit that breakdowns in communication are less likely to occur. Many of these concepts have been replicated in C/HAVs, such as keeping the driver in the loop regarding the performance and decisions of automation (Seppelt & Lee, 2019). Of most relevance is the work of Walch et al. (2017), Beller et al. (2013), and Verberne et al. (2012) who explore the design of cooperative interfaces for C/HAVs by applying the work of Klein et al. (2004, 2005). Their interface solutions bring together Klein et al.'s actionable concepts by demonstrating value in the application of performance information, capacity information, and convening on shared goals.

More recently, Naujoks et al. (2019) summarize previous research in C/HAV interfaces by providing 20 principles that C/HAV interfaces should be adhered to Scharf address safety concerns. These principles cover accidental mode changes, mode, state, interface position, grouping elements, time management, display characteristics, urgency, driver arousal, multimodality, directability, and consequences. Each principle outlines how major features of previous research in C/HAV interfaces should be implemented as part of the overall human–automation interaction to improve outcomes in C/HAVs. This book has addressed many of these factors, particularly how to communicate mode, state, and how to construct interfaces although some (e.g., driver arousal) are not addressed here. This is largely due to a focus on interaction and interface solutions, although the integration of psychophysical sensors to detect driver awareness is plausible.

The original contributions of this book and the C/HAV interaction presented and validated within this chapter is comprised of a demonstration of how user interaction can draw on concepts in bidirectional vocal exchange for raising SA while identifying and communicating essential information regarding the situation and vehicle state. The design is grounded in the theoretical concepts of distributed SA and joint activity and can adapt to multiple situations (i.e., flexible information transference for both emergency and nonemergency handover events). The validation stage of this chapter tests the performance of this novel interface with regards to the major factors identified previously in the book: preferences, vehicle control, usability, trust, communication, workload, and acceptance. The following sections outline how the book has contributed to this final design and how this design was physically constructed.

9.1.2 SUMMARY OF THE DESIGN PROCESS

Table 9.1 summarizes the main contributions of the previous chapters toward the final design. Throughout this book the automated assistant design timeline followed a detailed process of scoping, piloting, convergence, and testing. First, theory and current state of handover assistants were scoped (Chapter 2). This chapter outlined theory from both joint activity (JA) and DSA suggesting that a combination of these theoretical frameworks was necessary to address the cooperative nature of human–automation interaction while acknowledging the distributed nature of roles, agent capacity, and SA. Next, the AV domain was analyzed using cognitive work analysis to outline domain constraints, identify tasks, and assign functions for C/HAV interaction (Chapter 3), drawing on the theories outlined in Chapter 2. The resulting analysis showed how JA and DSA can be applied to communication within C/HAVs. Chapter 4 reviewed literature in human shift handover to generate possible strategies for interaction to take place in C/HAVs and evaluated these strategies using the principles of DSA (Chapter 4; Clark et al., 2019b). Four of these handover strategies were selected to be tested in a human–human handover pilot test, showing that questioning the outgoing driver was an effective strategy in communicating SA information (Chapter 5; Clark et al., 2019b). Building on these vocal interfaces, the addition of visual interfaces was then considered to ascertain which displays were more effective to display information. This study found that displays residing closer to the driver view were more relied on by drivers (Chapter 6; Clark et al., 2019c) and provided insights into the role of directability (outlined within the theoretical framework of JA). Using materials from previous chapters, such as the cognitive work analysis (CWA), experimental findings, and a description of the underpinning theory, the design process was then initiated by conducting design workshops with users and experts to generate potential solutions to improve C/HAV interaction (Chapters 7 and 8). Participants in these design workshops considered outcomes from previous chapters, linking the theoretical design recommendations throughout the book to a practical solution. Chapter 8 represents the combination of all previous findings, as these were the focus of the workshop when drafting the prototype concept. Chapter 9 takes the design from concept to physical prototype and then validates the

TABLE 9.1

Contributory Findings to Final Design and their Associated Chapters

Contributory Finding	Associated Chapters	Implementation
Is a multimodal interface[†]	2,3,8	Utilizing HUD, cluster, audio, and vocal information
Provides state information*[†]	2,3,8	HUD indicator and state light
Provides capacity information*[†]	2,3,8	Communicates to driver if dropout is expected
Adheres to distribution of situation awareness[†]	2,3,4	Raises situation awareness prior to driver taking control
Provides structure in protocol[†]	4,8	Control transfer steps and progression are consistent
Communicates vocally[†]	3,4,5,8	Assistant communicates to driver in full sentences
Provides contextual information[†]	4,5,8	User can query vehicle to provide information that suits context
Features bidirectional communication*	4,5,8	Assistant initiates a dialogue to aid in raising situation awareness
Allows users to question to raise SA*[†]	4,5,8	Assistant initiates a dialogue to aid in raising situation awareness
Display visual information close to road view[†]	3,6,8	Information displayed throughout journey is presented via HUD and cluster
Communicates what action is required*	7,8	Messages are communicated in a 'event/action' format to quickly inform the driver of what and why needs to be performed

Note: * = in line with joint activity

† = in line with distributed situation awareness

prototype by analyzing performance and subjective measures with users in a driving simulation.

9.1.3 OVERVIEW OF FINAL HANDOVER ASSISTANT DESIGN

Chapter 8 provides many elements that human factors experts have proposed as being capable of communicating the informational concepts explored throughout the book. This initial design was used as the foundation for the development of the final prototype within this chapter. Alterations to the initial design were necessary to ensure practicality, technical feasibility, and that all major concepts outlined in previous chapters were present.

Experts from the design with intent workshops in Chapter 8 created a blueprint for an automated assistant named 'Steeri'—an anthropomorphic vocal assistant that was able to communicate intentions, expectations, and information to the driver during all stages of a journey. Steeri could answer questions vocally to guide the driver through the automation cycle and raise SA. Virtual assistants like this have been found to be an effective tool in interfacing with users to ensure that users can navigate virtual environments and provide users with an adaptive and flexible interaction (Chérif & Lemoine, 2019; Cho, 2018; Parke et al., 2010). Steeri communicated to the driver via vocal and visual streams throughout the automation cycle and presented the driver with emotional faces based on the situation. Steeri was able to vocally communicate in full sentences to make sure that the situation was clearly communicated and ensured that 'capacity to perform' was made clear.

Steeri's features aim to improve human–automation interaction by optimizing trust, usability, and adhering to cooperative principles. To achieve this, Steeri's design encompasses many of the design recommendations generated throughout this book, summarized in Table 9.1. Steeri's design allowed for the driver to know what is to be expected and whether a handover is necessary (an extension of Verberne et al.'s (2012) findings in communicating capacity information in AVs). Adhering to concepts outlined by JA and distributed situation awareness (DSA), the interaction focused on alerting the driver, communicating state, communicating capacity, outlining future events, guiding the driver to what is required for physical transition, and allowed for a period of time for user querying. The following section outlines the process of going from design concept to prototype, as well as providing supporting literature where necessary.

9.1.4 The Development of Steeri

Steeri was designed and implemented using Visual Studio v16.1. Visual Studio mediated visual, auditory, and physical interaction. Figure 9.1 outlines Steeri's functions and interface elements at each stage of a journey. Lanes indicate what is happening and the displays that are shown at each time. For the head-up display (HUD), letters are given in relation to the icons presented in the following sections (Figures 9.2–9.4).

9.1.4.1 Visual Information—Head-Up Display

The first step toward implementing Steeri in a vehicle was to generate the visual aspects of the display. In line with findings from Chapters 6 and 8 suggesting that displays closer to the driver's front view is an effective strategy to communicate visual information during control handovers (Clark et al., 2019c), a HUD was chosen as the main visual modality. The HUD consisted of a banner at the bottom of the windscreen (see Figures 9.4 and 9.5 for examples). Visual elements of the HUD were rendered in Inkscape v1.0 to represent a female face, in line with current leading technology (e.g., Cortana, Alexa, and Siri). As the demographic of the virtual assistant was not controlled for in this experiment, this design decision was made on basis of familiarity and current market trends. This should not be received as being mandatory, rather, factory-ready handover assistants should be customizable to represent variations of race and gender to suit driver preferences and ameliorate the reinforcement of bias.

Steeri featured five facial states: happy, assisting, concerned, panicked, and questioning (presented in Figure 9.1). In line with previous research, the presentation of faces such as these may have positive effects on subjective preference and in some

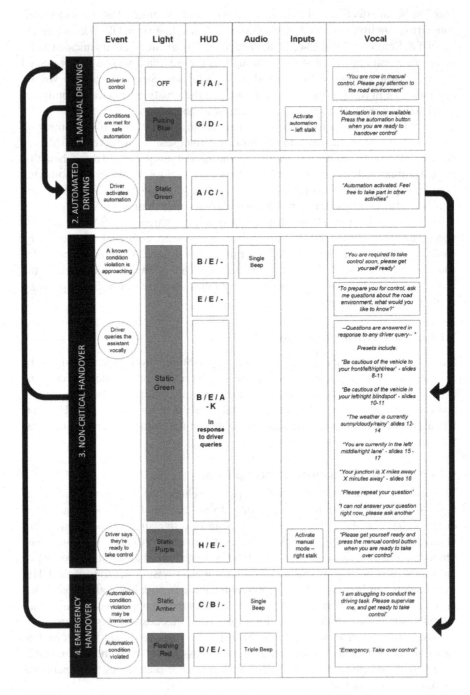

FIGURE 9.1 Final design flowchart for Steeri outlining modalities (lane), events, and information displayed during each control stage.

cases, improve task performance (Bass & Pak, 2012). Steeri's facial expressions were presented at varying stages of the journey either to indicate that automation is safely operational, information is being given, automation requires supervision, automation requires driver intervention, or opening a dialogue with the driver to ask questions. When action was required by the driver, icons representing the input required from the driver replaced the face icons to guide the driver toward correct physical operation. These are also presented in Figure 9.1, representing 'driver-in-control', 'activate automation mode', and 'activate manual mode'.

These face icons communicate intentions and capacity information as well as provide the driver with information regarding expected actions and vehicle state. To aid in the communication of vehicle state, these icons were accompanied by a 'state bar' to show which agent was in control at the current stage of the journey. The concept was drawn from the design in Chapter 8 and simplified to include three control states and the situations that can be replicated in a driving simulation: manual, supervised, and automated. The blue vehicle moved up and down the state bar to indicate the respective state. Arrows were added to this state bar to enhance the clarity of intentions and future actions/events. Figure 9.3 shows the state bar during manual, supervision, and automated mode, as well as the slides showing upcoming transitions to automation or manual mode.

The final element of the HUD was the addition of information icons when the handover assistant raised SA of the driver. The use of a HUD to raise SA was first discussed in Chapter 6 as an effective way of transmitting visual information, due to being situated close to the driver's default direction of gaze during the driving task. Workshop findings in Chapter 8 also suggested this as a way of transmitting visual SA information. These icons were solely presented to the driver in nonemergency handover situations and when the driver requested the respective information. The information types were sourced from the findings of Chapter 5—the most frequently

FIGURE 9.2 HUD icons displayed on the left—Steeri emotional faces and physical input requests.

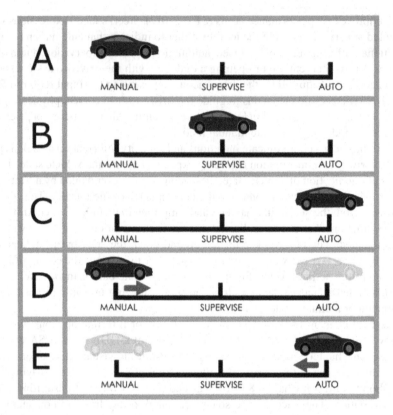

FIGURE 9.3 Image of the five instances for the HUD state bar, displayed in the center of the display.

requested information types during human–human handover in an automated vehicle (Clark et al., 2019a) were hazards, lane, exit distance/time, and weather. These icons are presented in Table 9.4.

Combinations of the icons presented in Figures 9.2–9.4 formed 19 different potential slides that could be presented to the driver during the journey, considering different urgency handover events. An example of a combined HUD slide is presented in Figure 9.5, and an example of what these slides look like from the driver view is provided in Figure 9.6. (See Appendix B for a full visual list of HUD slides.)

A 'state light' was presented to the driver to reiterate the current state of the vehicle. The state light is a feature found in current AV HMIs as a way of providing the driver with a quick indication of mode. This state light formed the 'control' comparison for experimental validation of Steeri. The state light was also included as part of Steeri's design to test how developments of current technology can improve human–automation interaction. Colors were replicated from the Cadillac Super Cruise (2020) with the addition of amber and magenta to communicate emergency and nonemergency handover situations, respectively. These colors are displayed in Figure 9.7. The state light was implemented into the vehicle via an Arduino Uno,

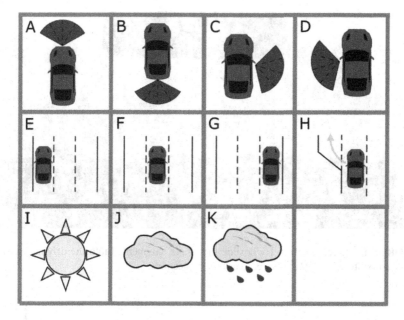

FIGURE 9.4 Situation awareness icons displayed to right of HUD during situation awareness raising visual information—state light.

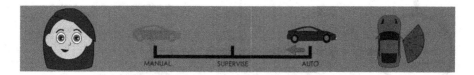

FIGURE 9.5 Example HUD slide displaying what could be shown prior to transfer of control: an emotional assistant (left), state and future state (middle), and information asked for (right; potential hazards).

interfaced with Visual Studio to display colors in line with the HUD banners outlined here.

9.1.4.2 Vocal and Audio Interfaces

Previous research indicates benefits of vocal interfaces over manual counterparts (Barón & Green, 2018). In support of this, workshop output from Chapter 8 shows Steeri being capable of having a bidirectional vocal dialogue with the driver to coordinate actions and make requests. Chapters 4 and 5 indicate that questioning the automation is an effective vocal strategy in raising SA. For technology, this concept is known as 'user querying', a method for users to request information in real time and has previously been demonstrated as an effective tool for user interaction (Lugano, 2017). User querying has the benefit of providing information to the driver

FIGURE 9.6 Example of an HUD slide in vehicle, augmented into the driving scenario. Slide shows Steeri prompting driver to ask their questions.

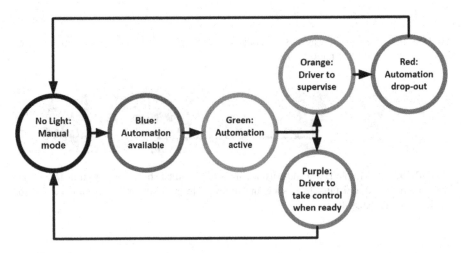

FIGURE 9.7 State light colors during each stage.

in response to what they require and can adapt to the current context, something a prescriptive method (such as a checklist) may be capable of addressing. This technique is common in shift-handover settings (Chapter 4; Clark et al., 2019b) and was found to be promising for C/HAV handover in Chapter 5 with the related condition improving a wide range of human–machine interaction metrics (Clark et al., 2019a).

Developing the vocal interaction for Steeri involved creating vocal messages that communicated information related to each stage of the journey. A single vocal message was presented to the driver each time the HUD banner changed. Steeri started each sentence with the event and then *the action* (e.g., 'You are required to take control soon, get yourself ready'). In doing so, drivers were made aware of the situation and what they were required to do following this event. Prior to nonemergency handover, Steeri initiated a dialogue to allow drivers to query the automated assistant regarding the situation. Questions could relate to the past, current, or future situation but were not restricted to any particular topic. Drivers could ask as many questions as they liked prior to handover and received answers via vocal prompts and the respective HUD slide was displayed (if available for that question). Once ready to take control, the driver vocally stated 'I am ready to take control', at which point, the automated assistant asked the driver to take control. This confirmation is essential for coordinating actions

Vocal interaction was coordinated via a 'Wizard of Oz' interface controlled by the lead reseracher. For raising SA, the program consisted of preset responses (such as 'watch out for the vehicle to your left') that were selected by the experimenter when a question was asked. A miscellaneous box also allowed the experimenter to answer questions that may differ from the preset answers. Preset answers on the Wizard of Oz interface were designed to represent the most popular questions from previous studies (Clark et al., 2018).

In addition to the vocal interfaces, when notified of the requirement to takeover control vocal prompts were presented alongside a single beep. When the automation dropped out, two beeps were presented to the driver.

9.1.4.3 Transferring Control

When required, drivers transferred control by pressing the end of an indicator stalk. Left for automation, right for manual. These were labeled 'A' and 'M' to guide the driver. When emergency handovers occurred, the experimenter deactivated automation from the control room.

9.2 METHOD

9.2.1 PARTICIPANTS

Participants were recruited through internal communications at the University of Southampton and through the university's website. The only requirements were to be above 18 years and hold a full UK driving license. Forty-six participants were recruited aged 22–75 (mean = 42.17, SD = 15.9), of which, 17 identified as female, 28 identified as male, and 1 identified as non-binary. Drivers had a mean of 22.9 years of driving experience (SD = 16.3) and a mean of 8,926 annual mileage (SD = 5,596). Ethical approval was given by the University of Southampton ethics committee (ERGO No. 52008).

9.2.2 EXPERIMENTAL CONDITIONS

Three display conditions were tested, each representing a different coordination level of Steeri. Table 9.1 outlines the entire automation cycle that takes place in a cyclical manner during each trial, along with vocal communication for both emergency and nonemergency handovers. Table 9.2 shows the conditions that were tested—a control condition consisting of a representation of a current level 2/3 vehicle interface

TABLE 9.2
Components of Steeri and Inclusion within Each Testable Condition

Interface Component	Control	Steeri	Steeri w/ Interaction
1. State light	✓	✓	✓
2. Basic auditory cues	✓		
3. Coordinative auditory cues		✓	✓
4. Head-up display		✓	✓
5. Situation awareness raising			✓

(Cadillac Supercruise; Cadillac, 2020); handover assistant named 'Steeri' providing more detailed head-up and vocal coordination; and an extension of the Steeri condition with the addition of a voice recognition system that allows for transactions of SA prior to a nonemergency handover (in which Figure 9.1 outlines). All five components, and by extension the interfaces displayed in Table 9.2, are illustrated in the following sections.

In the control condition, a single beep was presented when attention was required (upcoming handover) and two beeps for emergency 'take control now' states. The pink, orange, and red lights were accompanied with brief vocal prompts such as 'supervise' to help distinguish what the situation was.

Steeri conditions (with and without bidirectional interaction) kept beeps but extended the control's audio cues by providing more detailed vocal information in line with 'coordinated activity' theory base. This was centered around an emotional assistant that transmitted different vocal prompts related to the situation such as 'You are required to take control soon, please get yourself ready'. This was presented along with HUD elements to ensure multimodal communication of important information.

As raising SA had a large effect on how people viewed the handover process (Clark et al., 2019a), a separate condition was performed where user querying was removed; however, all other interface elements remained present. The control condition was adapted from the Cadillac Super Cruise (Cadillac, 2020), modified to distinguish between the two types of handover: emergency and nonemergency. Full details of interface elements and condition specifics are outlined in Section 9.2.2. It was hypothesized that Steeri, due to its careful and iterative HF design with both users and experts, would improve vehicle control following handover, usability, acceptance, effective communication, optimized trust, and optimized workload.

9.2.3 DESIGN

The experimental design consisted of three within-group trials, each consisting of one of the interfaces in a counterbalanced ABC design (control, Steeri, and Steeri with SA). In each trial, four vehicle-to-driver handovers were performed—two being nonemergency handovers and two being emergency handovers separated by 2–5 min of automated driving (2, 3, 4, and 5 min randomly assigned). Trials were counterbalanced systematically across the sample by participant number.

The independent variables were interaction condition and type of handover (non-emergency and emergency). Dependent variables were steering wheel angle, speed, subjective workload, trust, effective communication, acceptance, usability, and interface rank.

9.2.4 APPARATUS

The Southampton University Driving Simulator (SUDS) was used to simulate a highway environment. The simulator consists of a fixed base and a 135-degree front field view, with rear view and wing-mirror displays. The validity of testing driving behavior in driving simulators appears to be high, with behavioral measures in the simulator correlating highly with real-world measures (e.g., Eriksson et al., 2017).

The simulation was modeled in SCANeR studio (ver. 1.9), simulating a three-lane highway with moderate traffic density with overtaking turned off for nonhuman drivers (to allow for better trial comparisons). The scenario featured minimal bends in the road to ensure that steering wheel inputs were valid during transfers in control. A HUD was projected onto the main screen display using Visual Studio, and augmented features of the display were implemented using a black fill for projection. Arduino Unos were used to sync displays with the state light, and an external joystick was used to activate and deactivate automation from the control room.

9.2.5 PROCEDURE

Participants received an information sheet outlining their right to confidentiality, anonymity, and right to withdraw on agreeing to sign a consent form. Following this, they were given a brief introduction to level 3 and 4 AVs and talked through the experiment. When ready, participants were guided through the controls of the simulator and then ensured that they had the right seat and mirror adjustments. Participants were then guided through the basic elements of the handover procedure such as what the state lights represent and what an emergency and nonemergency handover would consist of.

Vehicle interaction was mediated via a Wizard-of-Oz interface following the flow diagram in Figure 9.1 (and the assigned HUD slides outlined in Appendix B), allowing the researcher to initiate the next stage of interaction (programmed in Visual Studio) and repeat messages where necessary. The researcher listened out for vocal responses and monitored driver engagement prior to moving onto the next stage.

Participants were asked to stay in the middle lane and to not overtake other vehicles. Participants practiced driving down the highway and transferring control between themselves and the vehicle. Once comfortable with the experiment, participants were introduced to their first of three interfaces. Interface orders were counterbalanced across participants.

Three trials were then conducted, one for each interface. Each trial was identical, except for changes to the handover interface and randomizations in timings and the order of emergency and nonemergency handovers. Each trial lasted for approximately 20 min. Participants started stationary surrounded by traffic on a standard three-lane highway. They then started driving, bringing up the speed to 70 miles per hour. Manual driving consisted of 1 min and did not have a corresponding state light after which the vehicle notified the driver that automation was available. Notifications varied based on which interface was present as outlined in Section 2.2, although every trial displayed a blue pulsing light. Participants then pressed the automation button on the end of the left stalk and were instructed to wait for the car to communicate that it is safe to relinquish control inputs. When safely automated, audio notifications were presented, and a green light indicated that automation was active—cues and HUD information varied by trial. Participants then took part in a secondary activity—a book or their smartphone brought by themselves.

Time between handovers varied as 2, 3, 4, or 5 min, with a 25% chance of each. Four handovers were performed: two emergency handovers and two nonemergency handovers. An emergency handover consisted of a supervision stage, where drivers were asked to put down their secondary task and look out to the road; this was communicated through a beep, audio cue, and an amber light—again, cues and HUD information varied by trial. After 1–10 s (randomly timed) automation then dropped out, presenting emergency audio cues dependent on the interface condition, two beeps also accompanied these cues, and a flashing red light was also displayed at this time. Drivers at this point were expected to take control of the inputs without the need to click the manual button. After 4 s, interfaces returned to their regular manual driving display.

Nonemergency handovers consisted of a notification that control is required and displayed a pulsing purple light. In the 'Steeri with interaction' trial, following this notification, another prompt was made asking the driver if they had any questions. During this period, the researcher was able to communicate to the driver via preset messages (speed, lane, weather, and exit distance/time) and was able to respond to any question that the driver asked the automated system by using a text-to-speech link from the control room. Once drivers were satisfied with the questions asked, the driver notified the automation that they were ready to take control and then the driver was prompted to press the manual control button.

Handovers and manual driving repeated until two of each handover type were complete (four in total), and then the trial was ceased. At the end of each trial, participants were asked to fill out questionnaires related to acceptance, usability, trust, effective communication, and workload. At the end of all three trials, participants were briefly interviewed as to which one was their favorite and why. They were asked to rank interfaces and provide any suggestions for improvement. Following this, they were thanked, debriefed, and paid £10 for their travel costs.

9.2.6 METHOD OF ANALYSIS

Speed was analyzed as a function of time using a linear regression for both nonemergency and emergency handovers with interface type specified as a predictor variable. Absolute steering wheel input was used to ensure that both left and right inputs were treated the same to represent steering control. For each participant, in each trial, for both emergency and nonemergency handovers, a mean steering wheel input value was calculated. These were then analyzed using a repeated measures ANOVA with both interface type and handover type specified as predictor variables. Subjective measures (interface preferences, workload, trust, communication, acceptance, and usability) were all analyzed using Friedman tests with Holm–Bonferroni corrected t-tests.

9.3 RESULTS

9.3.1 VEHICLE CONTROL MEASURES

For each interface condition and each handover type (emergency and nonemergency), the change in speed and steering behavior was analyzed. It was found that upon takeover, speed typically dropped by a small amount (approximately 1 mile per hour) with a dip at 2 s following handover to return to the speed of automation after 4 s. Drivers were not uniform in their speed change behavior. Standard deviations were plotted against time to illustrate the increase in spread in speed across participants. Linear regression analyses showed a strong effect of time on deviation from automation speed for both nonemergency, $F(1, 238) = 575$, $t = 23.98$, $p < .001$, $R = .71$, and emergency handovers, $F(1, 238) = 1,055$, $t = 32.49$, $p < .001$, $R^2 = .82$, indicating that human driver's speed increased in variance within participants as time progressed from the handover. Figures 9.8 and 9.9 show vehicle speed, speed standard deviation, and mean lateral velocity for 4 s following the driver regaining control of the vehicle for each interface condition.

Steering wheel inputs were analyzed post-handover to detect any vehicle control deficiencies across emergency and nonemergency/routine conditions, as well

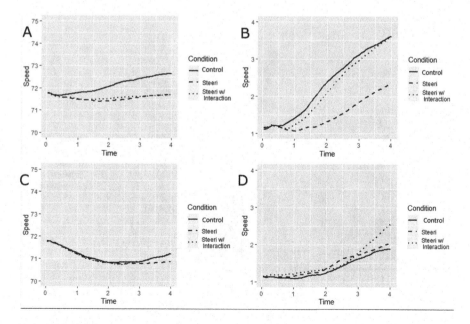

FIGURE 9.8 Four-line graphs to show true speed during nonemergency handover (A) and associated standard deviations (B), and true speed during nonemergency handover (C) and associated standard deviations (D). All plots are grouped by condition.

as between interface conditions. A repeated measures ANOVA, analyzing for an interaction between condition and handover type, found that there was a significant, although small, main effect of 'interface condition' on mean steering wheel input, $F(2, 224) = 3.93$, $p < .05$, $\eta p^2 = .034$, and no main effect of handover type on steering wheel input, $F(1, 224) = 1.86$, $p > .05$, $\eta p^2 = .008$. Further, there was no interaction between interface condition and handover type. Tukey's post-hoc tests found that mean control steering wheel input was statistically greater than both the Steeri condition and the Steeri with interaction condition indicating a potential reduction in control when vocal/head-up interfaces were not present.

9.3.2 Subjective Measures

Upon completion of the trial, participants ranked their favorite, intermediate, and least favorite designs. Figure 9.10 displays the frequency of ratings for each condition. The majority of participants (38 out of 46) ranked the control condition as their least favorite. Twenty-nine of the 46 ranked Steeri with the addition of SA as their favorite, with the majority of participants (31 out of 46) ranking Steeri with no SA as their intermediate option. Attributing ranks as being ordinal, a Friedman test showed a significant effect of design on ranks allocated. A post-hoc pairwise Wilcoxon test showed that these differences were present between control + 'Steeri' and control + 'Steeri with Interaction' conditions ($p < .001$).

Questionnaire data show improvements of Steeri being present across a wide range of measures. Figure 9.11 show box plots of ratings across acceptance, usability,

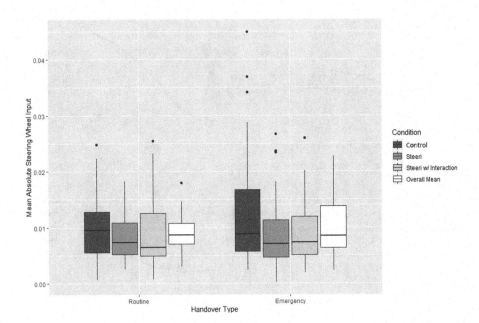

FIGURE 9.9 Box plots to show mean steering wheel inputs (absolute values) for emergency and nonemergency handovers between each interface condition.

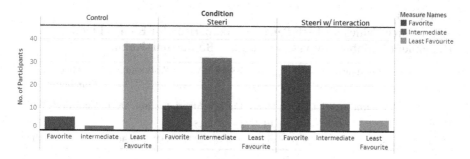

FIGURE 9.10   Bar charts to show frequencies for each rank of preference for each interface condition.

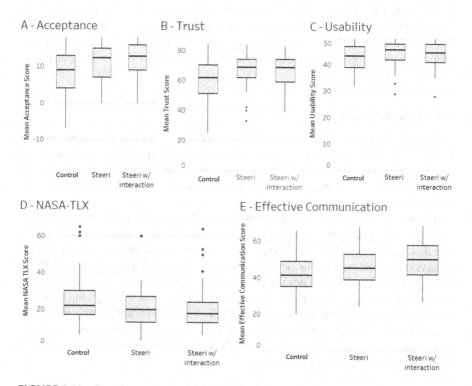

FIGURE 9.11   Box plots to show subjective ratings for acceptance, trust, usability, workload, and effective communication for each interface condition.

workload (NASA-TLX), and effective communication measures. Friedman tests showed a significant main effect of design on the subjective ratings of acceptance, Chi(2) = 11.96, $p < .01$; trust, Chi(2) = 8.74, $p < .05$; usability, chi(2) = 7.99, $p < .05$; workload, chi(2) = 6.37, $p < .05$; and effective communication, chi(2) = 21.06, $p < .001$. Post-hoc tests are displayed in Table 9.3.

TABLE 9.3
Significance Values for Each Pairwise Comparison for Each Subjective Dependent Variable. Alpha Corrections = Bonferroni–Holm.

	Acceptance		Trust		Usability		Workload		Effective Communication	
	Control	Steeri	Control	Steeri	Control	Steeri	Control	Steeri	Control	Steeri
Control	—	.02*	—	.01*	—	.54	—	.05*	—	.01*
Steeri w/inter	.01*	.1	.01*	.6	.54	.23	.04*	.46	.01*	.01*

* indicates $p < .05$.

Finally, a two-way dependent samples ANOVA found that individuals with a higher propensity to trust were more likely to trust the automated vehicle in all interface conditions, $F(1, 119) = 22.267$, $p < .001$. However, there was no interaction found between propensity to trust and interface condition, $F(1, 119) = 0.035$, $p > .05$. A Pearson's correlation showed propensity to trust and trusting AV interfaces were mildly correlated ($r = .39$, $p < .001$).

9.4 DISCUSSION

This study aimed to culminate outcomes explored by a wide range of studies evaluating interaction between driver and automation in level 3 and 4 AVs. The interfaces tested here include a current level 3 AV design and a novel interface designed using a HF design lifecycle aimed at improving a wide range of human–machine interaction outcomes.

The prototype automation assistant was developed as a culmination of previous chapters within this book, representing a novel approach to human–automation interaction—one that focuses on addressing driver awareness requirements and capacity to perform via two-way vocal interaction. The interaction design displayed real-time HUD information to coordinate actions to communicate system expectations to the driver and guided the driver through the steps required for handover while giving them the opportunity to gain more information prior to handover. This approach is unique as a fully designed handover assistant that can adapt to both emergency and nonemergency handover situations featuring context-specific information transfer has yet to be developed and tested in this domain.

9.4.1 OVERVIEW OF FINDINGS

By bringing together multiple aspects that are thought to improve specific outcomes of system performance such as distributed SA (Stanton et al., 2006, 2017b), effective communication (Clark, 1996; Klein et al., 2004, 2005), and shift handover in human teams (Clark et al., 2019b), the developed automation assistant demonstrates that an all-round, cooperative automated assistant with SA raising capabilities can improve operational safety, acceptance, trust, workload, and communication. Further, users greatly prefer the novel interface compared to a current leading manufacturer design, indicating that there is room for manufacturers to improve user-centered design in their next-generation automobiles. Usability was the only dependent variable not to display effects across interface conditions, explained either because of a ceiling effect of the usability measure as medians across all conditions are very close to the maximum score or in line with a previous study on levels of interaction in level 3/4 automated vehicles, reduced interaction may not have an effect on usability due to a reduction in user involvement (i.e., the lack of HMI may neither be usable nor unusable, depending on user interpretation).

Findings also suggest that SA information, if presented in the right way, can be added without detracting from user experience—an issue encountered by Clark et al. (2019a) as part of vocal communication trials. Comparisons between the presence and absence of user querying prior to nonemergency handovers in this novel interface show to have no effect on vehicle control and subjective outcomes except for ratings in effective communication. This is likely due to bidirectional factor of this interface, as the interface actively listens to the driver during the journey. This is promising, as Chapter 5 demonstrated that the addition of SA assistants could have negative effects on desired outcomes such as usability and acceptance (see Section 5.3.3—checklist conditions). It follows that interaction should be considered from multiple perspectives while addressing user requirements and suggestions throughout the design process.

This study shows that user querying is an effective tool in level 3 and 4 automated vehicles, supporting the findings of Chapter 5, particularly as a way of raising SA in future vehicles. This finding provides further evidence for the effectiveness of the approaches explored by human–machine interaction researchers previously (Chérif & Lemoine, 2019; Cho, 2018; Parke et al., 2010). Realistically, this approach will have to be balanced with safety concerns, as automated interfaces may need to display safety-critical information as a default and provide additional information should the driver require it. Users preferred user querying to be present when asked to rate their preferred interface, but user feedback indicated this was partly due to reducing boredom and countering fatigue effects.

9.4.2 Relevance to Theory

In this book, the combination of DSA and JA (Klein et al., 2004, 2005; Stanton et al., 2006, 2017b) has shown to be beneficial for human–automation interaction. These frameworks have been influential in identifying appropriate methods for raising SA and giving both automation and driver opportunities to evaluate the state, phase, and capacity of the system results. User-querying synergizes with DSA nicely, as the user can access information that they may need for their own mental model that the AV may not readily provide. By providing multimodal awareness information, by facilitating transactions in SA through dialogue, while adhering to their roles as both driver and automated assistant, the results provided in this chapter demonstrate that DSA is a suitable framework for the AV domain. DSA provides the 'how' whereas JA provides the 'what'. Communicating state, phases, agent capacity, goals, and utilizing coordination devices has shown to be effective. For example, providing the driver with uncertainty information (extension of Verberne et al.'s (2012) work on uncertainty information in AVs) has been further reaffirmed. By providing the driver with upcoming state, actions, and providing readily accessible information regarding this (e.g., HUD on mode and facial icons), a well-rounded automated assistant solution can be developed.

Notably, neither this study nor this book has attempted to provide a granular investigation into which individual tenets of each theoretical approach improve outcomes nor does it address each tenet equally. For example, the tenets of 'agreement to collaborate' and 'common ground' are perhaps the least explored in this interaction. It is assumed, from an experimental perspective, that both agents are working together to perform the task. In real-world environments, this may not be the case. A driver of a C/HAV may not fully understand the benefits and drawbacks of automation, or fully understand how the vehicle functions, and so may fight the AV for control throughout the automation cycle. Common ground, as a further example, builds over time. However, in this study, the use of confirmations (e.g., 'you are now in control') allow communication to be grounded, and a clearer understanding of what has occurred is achieved. The role of DSA and JA in C/HAV interaction is explored further in Chapter 10.

The anthropomorphic nature of Steeri, although not a central concern of this book, provides a platform for users to interact with in a natural manner. This design feature was a natural solution to the foundational work presented to human factors experts in Chapter 8. However, neither the chapter nor the book has directly compared the presence and absence of an anthropomorphic automation assistant. Regardless, the feedback for Steeri is greatly positive with participants indicating that Steeri was more trustworthy and better at communicating with them compared to the control condition. These findings may support recent research showing that anthropomorphism in AVs is beneficial for user interaction (Waytz et al., 2014). The findings in Chapter 9 may be partly attributable to the anthropomorphic nature of Steeri; however, this effect remains unmeasured in this book.

Aside from DSA and JA, aspects of this interaction that may be of interest for further research include the role of boredom (perhaps objectively measured as low workload) and fatigue in relation to interfaces that promote more interaction compared to those that require less human participation. It follows that interfaces that require human communication at regular intervals could contribute to aspects of workload in line with malleable attentional resource theory (MART; Young & Stanton, 2002a, 2002b, 2007b)—the premise that underload can lead to a reduction in attentional capacity throughout a task.

9.4.3 Relevance to CWA

During the design process, the automated assistant was tailored toward the CWA presented in Chapter 3. First, the final prototype addresses the main factors outlined in the CWA abstraction hierarchy of Chapter 3. In particular, the solution provides a way of interacting via visual, vocal, audio, and haptic modalities, each addressing the key physical objects identified within the abstraction hierarchy (Figures 3.1–3.3). By addressing these modalities, each of the values and priorities identified outlined from DSA and JA have been addressed. These are outlined in Table 9.4.

TABLE 9.4
CWA Purpose-Related Functions and Associated Design Characteristics

Purpose-Related Functions	Design Characteristics
Provides information to driver	Accessible information via HUD, HDD, vocal interaction, audio messages, and state light
Creates communication link	Two-way interaction via vocal communication with visual and audio feedback. Physical buttons provide control transfer inputs
Facilitates control transition communication	Preset information transfer prior to handover
Adapts communication to driver requirements	Flexible user interaction via user querying
Assesses system safety	Capacity information and user querying

Table 9.4 shows that the primary purpose-related functions of the CWA are directly addressed by features within Steeri's design. Through a combination of multiple modalities, two-way interaction and the ability to direct the other agent, the functional purpose ('to facilitate effective communication') can be achieved.

9.4.4 RELEVANCE TO HUMAN–HUMAN STRATEGIES

The strategies identified in Chapter 5 (shift-handover strategy literature review) show that an overly structured approach to handover is at risk of omitting contextual factors involved in high-risk domain environments (Anderson et al., 2015; Bulfone et al., 2012; Cohen et al., 2012; Drach-Zahavy et al., 2015; Staggers & Blaz, 2013). The main strategies include vocal communication, bidirectional exchange, use of technology, adaptation, compatible mental model, contextual handover, and clarifying control (Clark et al., 2019b). Steeri provides a solution to this by providing preset information categories for drivers to access while enabling the driver to access contextual information as they see fit. The ability for both agents to send messages to the other vocally allows for immediate and efficient communication as well as allowing both agents to challenge the other during the automated cycle. As Steeri outperforms current human–automation interaction (the control condition) in the findings of Chapter 9, the results presented in this chapter indicate that learning from human–human teams may be beneficial for informing human–automation communication. As technology develops toward more anthropomorphic and voice recognition capabilities, these findings may become more relevant with time.

9.4.5 LIMITATIONS

User querying in this study was coordinated via vocal interaction from both parties. Applying this to a manufactured vehicle may require further attention, as this simulation did not capture the in-vehicle acoustics of a real-life environment. Further, drivers took part in this simulation with no passengers or music. With these factors present, user querying in this way may need to be considered via another interface modality. User querying as a way of raising SA has some drawbacks including user input may not be understood, user input may not be possible to address, and user priority not necessarily correlating with actual priority with regards to operational safety. However, users are provided with control over what information they want to receive, when information is requested, and are able to access contextual information related to the environment. Further, the user having control improves acceptance, meaning that they are more likely to use automated features as a result. The design presented here is discussed further in Chapter 10 with regards to how it addresses book outcomes.

9.4.6 CONCLUSION

This chapter represents the primary practical outcome for this book: a situationally sensitive, bidirectionally communicative, and awareness-focused automation assistant. The core concepts for modern human–automation interaction with automated vehicles were tested together in an experimental paradigm comparing current level 3 AV HMI with the generated prototype named 'Steeri'. The chapter's findings have shown that there are multiple benefits to implementing the approach outlined within this book including higher user preference, better acceptance, optimized workload, increased trust, and better communication. The validation presented in this chapter culminates all previous works and confirms that designers can make use of these HMI strategies to improve performance in C/HAVs. The next chapter concludes the findings of previous chapters to summarize design recommendations and discusses them in line with the book's research outcomes and hypotheses.

10 Conclusions

10.1 INTRODUCTION

Although cooperative interfaces have been explored previously in conditionally and highly automated vehicles (C/HAVs), this book provides a new and promising perspective in how cooperative communication can be approached in the domain of automated vehicles (AVs). The aim of this book is to design a novel interaction in AVs with a view of addressing the vulnerabilities introduced by semi-AVs such as the deterioration of situation awareness (SA), reductions in safe control of the vehicle and uncalibrated trust while addressing workload, usability, and acceptance. The book applies the theoretical concepts of joint activity (JA), which have been shown to be effective in automation design and applied following the theory of distributed situation awareness (DSA), a modern approach to addressing SA where agents' roles and responsibilities vary. This was achieved through following a four-step approach: scope, pilot, design, and test.

The following sections provide a summary of key findings, an evaluation of the method, and provide implications for future research and manufacturing activities in the domain.

10.2 SUMMARY OF FINDINGS

10.2.1 RESEARCH OUTCOMES

The chapters found within this book collectively address four research outcomes related to interactions in AVs: one primary outcome and three secondary outcomes. Findings are summarized within the following sections with regards to their associated outcome.

10.2.1.1 Primary Outcome—To Provide an HMI Design Solution That Improves Coordination Between Driver and Automation in Level 3 and 4 AVs During All Phases of a Journey

Each chapter within this book provides recommendations for improving communication in C/HAV assistants. Chapter 3 identifies SA, trust, safety, efficiency, usability, and acceptance as being key to ensuring that communication is effective. Through these values, many findings within this book contribute to this outcome. Most notably, the use of bidirectional communication (primarily via a vocal stream) was a major recurrent theme that formed the core handover communication in the final design. Chapter 2 describes how both parties should communicate intentions, status, and phases while providing transactions in SA when the counterparty requires them. Chapter 5 demonstrates the effectiveness of this vocal method in a driving simulator with two human drivers and Chapter 9 validates this method in a driving simulator within an human–machine interface (HMI) prototype. While developing concepts of effective communication (Clark, 1996; Klein et al., 2004, 2005) and developing an interaction addressing the presence of DSA (Stanton et al., 2006, 2017b), the vocal interaction featured in this book contributes toward optimized communication and serves as a promising foundation for implementing vocal interaction in automated assistants.

The final design presented in Chapter 9 features many additional concepts featured within previous chapters such as a coordinative head-up display (HDU) (inspired by the eye-tracking findings of Chapter 6) that communicates the phase, state, and capacity information. The practical aspects of the design were generated through recommendations provided by users and experts in Chapters 7 and 8.

Previous work in C/HAV interaction fails to address the breadth of the domain within singular studies. Alone, the findings discussed in this section are influential, but they are not all provided as original contributions within this book. Rather, this book's main contribution is a sequence of steps for developing new HMI, an interaction solution that features fully integrated concepts, and a way of addressing multiple design values in C/HAV interaction, validated via controlled experimentation. The final design is presented in detail within Section 9.2.2, with additional materials available in Appendix B.

10.2.1.2   Secondary Outcome 1—Provide Insight Into How Communicative Concepts and Distributed Situation Awareness Can Be Applied to Level 3 and 4 AV HMI Design

The chapters within this book contribute unique findings on how communicative concepts can be applied to C/HAVs. Chapter 6 addresses the role of directability, a key concept of JA (Klein et al., 2004, 2005), within C/HAV interaction. It provides a case for the use of vocal communication to guide the driver's gaze toward key areas of interest. Chapter 9 demonstrates that communicating state, capacity, and phase coordination through vocal and visual displays which may contribute to a wide range of positive outcomes for interaction. For DSA, of great value are the findings in Chapter 4, identifying and discussing 19 handover strategies in shift handover and how well they address the tenets of DSA. Not only can these strategies improve SA in AVs, they can also be applied to any domain that requires the continuity of tasks from one agent to another. These findings show that vocal communication and bidirectional feedback, due to their immediate and mutual nature, adhere to DSA principles and allow agents to access information that they require in real time. The book contributes to further discussion in the role of coordinated activity in human–machine teams and provides examples of how they can be applied within human–machine interfaces (HMIs) to improve interaction.

10.2.1.3   Secondary Outcome 2—Demonstrate How a Four-Step Approach to Human Factors Design Can Be Used to Address Multiple Domain Values

In summary, this book followed four steps to develop a design solution for the vulnerabilities introduced by C/HAVs—scope, pilot, design, and test. Scoping allowed for key theoretical concepts and the current state of the domain to set the foundations for the design stage. During this stage a domain and social analysis identified the structure and phases present in C/HAV operation. A review of handover strategies in shiftwork allowed for potential solutions to be drawn from other domains. Piloting these strategies and exploring the effectiveness of vocal and visual displays in coordinating activities in C/HAVs allowed for the book to identify what concepts should be taken to the design stage. In the design stage, these concepts were refined and discussed with users and experts using well-cited methodological techniques. Finally, testing allowed for a potential solution to be experimentally compared to current AV HMI.

The book, arranged in this way, provides research phases and examples of how a designer could approach novel design generation that considers the target user and addresses theoretical and practical issues within the domain. The final design's positive outcomes presented in Chapter 9 demonstrate the success of this four-step method.

10.2.1.4   Secondary Outcome 3—Provide Findings That Show How Driver Demographics May Affect Driver Requirements for C/HAV Interactions

The main findings presented in this book are supplemented by additional findings on how interaction varies between different user groups and demographics. Chapter 6 provides an investigation into visual gaze and finds no convincing difference in visual gaze behavior during the handover task between age groups, gender, time out-of-the-loop, and car ownership. Chapter 7 explores in more detail the difference in requirements between learner, intermediate, and advanced drivers. These findings indicate that learner and intermediate drivers prefer multimodal interfaces that guide the driver through the states and actions required during the journey. Advanced drivers, however, prefer less information to be displayed and prefer more autonomy on how the automation performs (Clark et al., 2020).

10.3   EVALUATION OF THE RESEARCH APPROACH

This book uses a variety of methods that are well documented in human factors (Stanton et al., 2017a). Cognitive work analysis (CWA; Vicente, 1999) is a well-cited method of mapping a domain and provides researchers and designers with a foundation in domain structure, activities, training requirements, and social organization. For C/HAVs, the social and cooperation analysis is particularly useful, as function allocation is regarded as being crucial for operational safety where automation is present (Fuld, 2000). The review process in Chapter 4 followed a thorough approach of systematically reviewing literature, using a range of search terms. This chapter outlines the criteria required for review and considers 799 research items. As a result of this method, the strategies identified are well represented in a range of domains. Systematic reviews constructed in this way are effective at retrieving and appraising current literature around a specific topic, like that found in Chapter 4 (Møller & Myles, 2016).

Three driving simulations were conducted in this book, although driving simulations are not fully representative of the complex and dynamic nature of real-world environments, human performance in driving simulations have been found to correlate highly with real-world driving tasks (Eriksson et al., 2017). Evaluations of driving performance involves a combination of behavioral (lateral and longitudinal control) visual gaze (Clark et al., 2019c) and subjective measures of workload (NASA-TLX; Hart & Staveland, 1988), usability (System Usability Scale; Brooke, 1996), acceptance (System Acceptance Scale, van der Laan et al., 1997), trust (Yamagishi & Yamagishi, 1994), and effective communication (Blake Group, 2020).

Qualitative findings are presented in Chapter 5 regarding driver perceptions of vocal strategies. The mixed method approach in this chapter provides findings that are both statistically evaluated, with the addition of rich data illustrating the user experience. This method is beneficial in other domains, such as healthcare (Östlund et al., 2011), and provides valuable findings for developing designs in line with driver requirements. The metrics selected in these experiments provide statistical breadth and ensure that design values identified in Chapter 3 are measured in line with the functional purpose.

The design stage features participatory workshops with target users and design with intent (DwI) (Lockton & Stanton, 2010) workshops with human factors experts. Participatory workshops are regarded as essential in human factors design (Sanders, 2003), and Chapter 7 presents detailed design recommendations that supports the benefit of this approach. In Chapter 8, DwI was used to generate recommendations in line with 101 design considerations. DwI is well cited; however, applications of this method to target domains are not currently well documented. It is hoped that this book demonstrates how DwI is influential in bridging preliminary findings and design recommendations to testable and fully integrated prototypes.

This book showcases a broad range of methods in each stage of scoping, piloting, designing, and testing. This book provides researchers and manufacturers with example methods for progressing through their design pathway. At each stage, methods and metrics can be replaced in line with the purpose of the research and design. For example, during the scoping stage a researcher may find it more appropriate to use a method such as hierarchical task analysis to gain insight into the target domain. For experimental work, a researcher may opt for on-road studies to ensure ecological validity or rely on qualitative research for more richness in their data. Approaches vary, and trade-offs are necessary. For further discussion on human factors methods, Stanton et al. (2017a) provide an in-depth outline for sociotechnical analysis and methods of analysis for developing solutions in human factors, irrespective of the target domain.

10.4 IMPLICATIONS OF RESEARCH

10.4.1 Notable Practical Contributions

This book consists of a thorough investigation into a breadth of issues in C/HAV interaction. The chapters within this book contribute individually and collectively toward literature in interaction development and testing. This book's primary outcome is practical in nature—the development of a fully integrated, C/HAV interaction design that aims to ameliorate vulnerabilities introduced by the transfer of control. Practical outcomes are numerous throughout the book for the C/HAV domain and other domains involving handover. Practical solutions are provided in all chapters, with many concepts remaining undiscussed and unexplored due to constraints in resources. Nine notable contributions are outlined in Table 10.1 and discussed in turn.

TABLE 10.1
Notable Contributions of the Book

Notable Contribution	Chapter Number(s)
Optimize outcomes in response to trade-offs	5, 6, 9
Allow drivers to access information	2, 4, 5, 8, 9
Integrate vocal communication	2, 3, 4, 5, 6, 7, 8, 9
Prioritize visual displays close to road view	6, 9
Communicate event and action required	2, 5, 9
Consider interaction at all stages of journey	3, 7, 8, 9
Allow driver to customize interaction	5, 8
Informing driver training requirements	2, 3, 4, 5, 6, 7, 8, 9
Application of human-team strategies	4, 5, 6, 9

10.4.1.1 Optimize Outcomes in Response to Trade-offs

Previous research into AV interaction fails to acknowledge the complexity of the tasks being performed in AV operation. Due to traditional approaches to experimental procedures and hypothesis testing, previous research into optimizing interaction is at a risk of being reductionist, as studies typically address improvements in relation to a single outcome. Currently, the research community in C/HAVs is at risk of promoting design solutions that may be beneficial in one regard but when integrated into a final product with additional features may negatively impact other important interaction outcomes. Results from the first experimental paradigm within this book (Chapter 5, human–human handover in an AV) showed that some interaction designs may improve safety but in turn could detract from usability. Lessons learned early on in this book show that multiple outcomes must be addressed, as having low usability may result in the system not being used at all. Chapter 5 showed that presenting minimal amounts of information to the driver inferred benefits such as a reduction in frustration and a decrease in handover time.

Chapter 5 concludes that prescriptive methods were a contributory factor to a decrease in acceptance and usability, whereas a flexible, user-querying method provided an optimized solution to the dependent variables measured. If alternative solutions had not been tested, researchers could have assumed that this decrease in acceptance and usability is solely attributable to the increase in situational information being received. By testing differing design concepts, this type 1 error could be avoided.

Future research into C/HAV interaction should measure interacting variables to capture the 'big picture' to understand which trade-offs should be made and provide solutions that can improve safety, as well as improving ratings such as acceptance and usability. Multiple designs should be tested, as the variation in interaction styles trialed throughout the book allowed for findings to indicate the most influential design solution.

This book succeeds in demonstrating how a rigorous and stepwise approach to interaction design can lead to improvements in a broad range of outcomes as the data of Chapter 9 show that as long as design drawbacks are identified early on in the design lifecycle, there is enough resources available to steer the design procedure down the right path. Something that human factors specialists have noted for many years.

10.4.1.2 Allow Drivers to Access Information

As previous automation domains have faced similar challenges to that of the AV domain, it is tempting for researchers in automation to draw on prescriptive methods found in domains such as aviation. Issues arise with this analogy, for example, drivers are not trained to a similar standard, procedures are not as easily instilled in the general public as they are in a work force, and purchasing/uptake of C/HAVs will be directly related to the user experience. Further, other high-risk domains challenge this approach and identify contextual information to be of paramount importance when conducting shift handovers. Access to contextual information includes any information that may be atypical for the average situation and may not be included in prescriptive methods. For C/HAVs this could include roadworks, future weather conditions, unidentifiable vehicle faults, and emergency procedures of the current roadway among many other situational states.

Having the driver in control of what information was given to them, requested in real time, was found to be beneficial during testing stages and shows that they can access is deemed as preferable and more adaptable to the situation. The dynamic interaction between driver and vehicle improved many outcomes while providing the driver with information that they deemed to be important at that given moment. Standardized information could still be implemented to instill a learning effect especially if information is essential in all scenarios (e.g., hazards). However, more work is required regarding what and how essential information should be combined.

10.4.1.3 Integrate Vocal Communication

Vocal communication is immediate and can be conducted while attending to manual tasks. Such communication allows for the driver to interact with the vehicle while performing secondary tasks (e.g., reading a book) or, depending on local regulations, while manually driving the vehicle. Vocal communication was found to be beneficial to those in shift-work domains and C/HAV operation alike. Vocal interfaces have been explored in great detail for use in AVs; however, this book demonstrates the modality's capability for managing high-risk situations in C/HAV operation.

10.4.1.4 Prioritize Visual Displays Close to Road View

Mainstream methods of information display in current C/HAVs feature a center console to conduct and manage tasks. Having such information in this area may mean that users are not directing attention toward the area naturally, as their attention is likely going to be paid toward the road environment ahead. It follows that drivers in Chapter 6 utilized displays closer to the road view to receive situation information compared to the center console. Going forward research should explore how augmented displays and HDUs can communicate critical information.

10.4.1.5   Communicate Event and Action Required

Throughout the book (e.g., Chapters 5, 7, and 9), users expressed their desire for the vehicle to express 'why' a handover is required. Alerts such as singular beeps or tones are vague and can instill worry in the driver, even if the situation is benign. It follows that for all events that may occur, the vehicle is able to relay quickly and effectively the reasons behind the interaction occurring and the expectations the vehicle has for the driver. These concepts were explored in detail in Chapter 2, as JA indicates that directability plays a role in coordinating activity. Alerts and notifications can make use of this format as the basis of message

10.4.1.6   Consider Interaction at All Stages of Journey

Previous research not only presented reductionist views on C/HAV interaction with regards to experimental variables but also typically explored vulnerabilities and information transfer during the handover. As is demonstrated in Chapter 4, other domains appreciate the role that pre- and post-handover activities have with task performance. This book provides solutions to C/HAV interaction at each stage of the journey and advises future research to also consider temporal factors.

10.4.1.7   Allow Driver to Customize Interaction

Customization was found to be a recommended strategy within the experimental findings, user workshops, and the expert workshop within this book. Data show that experiences, preferences, and suitability for specific demographic traits can all be addressed by integrating an interaction design that can be modified to include certain features. In Chapter 7, the idea of customizing for specific journeys was introduced. An example is a journey made with children passengers, so that more appropriate communication modalities can be selected if required. This form of customization could also benefit those with hearing and other such sensory difficulties. Another form of customization is the tailoring of information displayed during handover interaction. For example, one user may not find fuel to be important, whereas another may prefer it to be delivered prior to taking back control. In this way, a user schemata and experience can be addressed with the information that the user requires via customizable features.

10.4.1.8   Informing Driver Training Requirements

Collectively, the chapters within this book lead to the importance of extending the current taxonomy of the automated cycle to include 'awareness' and feature more driver-focused terminology to inform driver requirements during training. For those learning to drive for the first time, it could be important to understand how C/HAVs could operate in the future, particularly if a mix of both emergency and nonemergency handovers are required. A proposed approach to this is 'the three A's':

Attention—Ensure that all secondary tasks are set aside including discussions with other passengers, interaction with devices, or other objects.
Awareness—Eyes out on the road, address hazards and vehicle state—e.g., weather, speed, and other vehicles.

Action—Assume a comfortable position, ensure contact with all control inputs (accelerator and steering wheel), and initiate the transfer of control when ready.

The three A's could be adapted to address both emergency and nonemergency handover events. The role of time constraints may require the awareness stage to be shortened or delayed as quick action may be safety critical.

10.4.1.9   Learning From Human Teams

This book represents one of the most thorough investigations into how modern technological capabilities can be designed with natural communication in mind. The theoretical and practical underpinnings of this research is that of human-shift work. Steeri was informed primarily by research into how handover is conducted within domains such as healthcare by integrating strategies such as questioning, a two-way dialogue, and vocal communication. This design outcome indicates that automation capabilities in preexisting or new domains could learn from human teams when informing their interaction design process.

10.4.2   Notable Theoretical Contributions

This book demonstrates that the theory of DSA and JA can be successfully applied to C/HAV interaction design together. DSA—the premise that SA is distributed across agents and element in a system (Stanton et al., 2006, 2017b)—is used to explore and design how human–automation interaction should take place in AVs. DSA's scope does not include specific guidance as to what should be communicated during C/HAV interaction. JA was selected to guide the content of communication due to its task-centered approach to human–machine interaction (Klein et al., 2004, 2005). JA focuses on communicative concepts that give agents a better understanding of how their counterpart is performing and a model for collaborating effectively during shared tasks. This book provides valuable contributions toward progressing the theoretical framework of JA into a distributed form while showing that SA is a collaborative process that builds over time. Notable theoretical advances are outlined in the following.

10.4.2.1   User Querying—Addressing SA Requirements and Grounded Communication

The use of bidirectional communication has multiple purposes. Ensuring that both agents can send messages within shift-work handover ensures that information can be challenged, reaffirmed, or corrected (Drach-Zahavy & Hadid, 2015; Parke & Kanki, 2008; Rayo et al., 2014). This form of communication also addresses the issue of humans and machines perceiving the situation differently. DSA states that SA should factor in the difference in processing capabilities between humans and machines (Stanton et al., 2006, 2017b). However, it is difficult for a human or machine to understand the SA requirements of their counterpart. Bidirectional communication in this way allows each agent to access information that they may require from the other agent to raise their own perception of the situation.

10.4.2.2 New Insights Into a Virtual Copilot

The findings within this book provide further evidence that designing automation to be a copilot, rather than a tool, is beneficial for user interaction (extension of Eriksson & Stanton, 2017c; Walch et al., 2017). Following a scripted approach appears to detract from user experience. Whereas allowing drivers to interact with the vehicle more freely improves interaction in a variety of ways (Clark et al., 2019a). The communication of state, capacity, and phase information has been shown to provide positive outcomes for human–automation interaction in AVs while the use of two-way interaction allows both driver and automation to relay information and address misunderstandings in real time. As such, the concepts outlined by Klein et al. (2004, 2005) and the work of Clark (1996) have been verified to be beneficial to AV design and remain a promising line of enquiry for automation researchers interested in applying the concepts of human–human communication to automated systems.

10.4.2.3 Supporting Transactions in SA

In line with DSA (Stanton et al., 2006, 2017b), this book contributes toward how transactions (Sorensen & Stanton, 2016) should be performed in human–automation systems. A mix of providing the user with essential information (such as a notification or warning) and providing a user querying service could solve the 'how much is enough' situation that many HMI designers encounter. The user can be provided with safety-critical information and can then access further information should they deem it to be necessary. In this way, the user remains in control of a proportion of the information they receive, although designers can integrate safety information should this be required. It remains a challenge in understanding what is essential information. However, trials such as that conducted in Chapter 5 could indicate what drivers naturally require during the shift handover. When combined with methods tailored towards addressing safety concerns (such as CWA and DwI), solutions for addressing this complexity could be readily acquired.

10.4.2.4 Demonstrating the Perceptual Cycle Model in Action

The perceptual cycle model, introduced in Chapter 4 and discussed in Chapter 7, represents the cycle of world (real-time environmental state), action (the decisions and behaviors exhibited by the system), and schemata (the mental models used to comprehend perception information), each directing and modifying one another in a cyclical fashion (Banks et al., 2018; Neisser, 1976; Plant & Stanton, 2012, 2013; Revell et al., 2020). As environmental states change, actions are made informed by schemata, and schemata are developed in turn. This cycle represents the foundational aspects of DSA as each agent within a system has their own schemata (especially between human and machine agents) and perception of the environmental state therefore leading to noncomparable forms of SA between agents (Stanton et al., 2006, 2017b).

This book presents findings that show that improving the link between world and action can have benefits for interactions in C/HAVs. Providing a way of raising SA and guiding the driver toward appropriate actions features greatly in Steeri's design.

Further, the ability for the driver to access information via user querying allows the driver to address their own schemata to gain a better awareness of the situation. Chapter 7 shows how users with varying experience may interact differently with C/HAVs. Therefore, this book provides further evidence that the perceptual cycle model and DSA are effective models for approaching the issue of interaction with C/HAVs.

10.4.2.5 The Role of Malleable Attentional Resources Theory in C/HAVs

Malleable attentional resources theory (MART), the premise that attentional resources synchronize with workload, suggests that workload should be optimized, not reduced (Young & Stanton, 2002a, 2002b, 2007b). In this book, multiple instances of boredom and frustration were reported for both checklist conditions tested in Chapter 5 and the control condition in Chapter 9. A dialogue-based system that allows the driver to interact with the AV may allow the driver to keep up higher levels of attentional capacity due to an increased requirement to interact. It follows that drivers may develop greater levels of SA and feel more integrated with the system. MART in this way could provide insights into these findings and be an effective tool in optimizing driver interaction.

10.4.3 Notable Methodological Contributions

This book provides a methodological framework for developing human factors designs in new domains. The process of scope, pilot, design, and test has provided the design outcome with a solid theoretical and practical foundation and ensures that multiple outcomes and avenues are considered while addressing user requirements throughout. Further contributions are presented for potential applications of CWA, driving simulations, participatory workshops, and the 'design with intent' method of design generation. For driving simulations, measurable outcomes are showcased for measuring the quality of interactions between driver and automation in C/HAVs. Figure 10.1 outlines the final outcomes for each chapter and how they relate to the final design's development.

Figure 10.1 and the chapters as a collective can provide a useful resource for the design lifecycle for C/HAV interaction and may also inform human-interaction design in other domains. The approach is particularly influential as it covers a broad range of issues human–automation interaction faces. These stages can be implemented into new and current domains to factor in theory, current practice, learn from other domains, stepwise testing of concepts, and contributions of both users and experts in the design process. This approach to human–automation interaction allows the designer to gain rich data and broad recommendations that can be refined into a single concept.

10.5 FUTURE WORK

10.5.1 The Balance of Priority and Supplementary Situation Awareness Information

The findings presented in Chapters 5 and 7 show the range of information types prioritized and requested by drivers. Due to the book following the thread of user

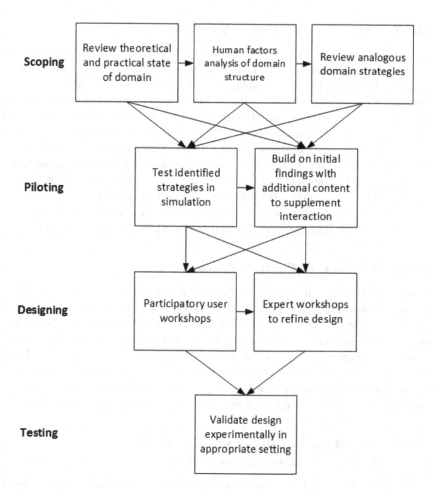

FIGURE 10.1 Redrawing of Figure 1.1 to present general approach and research framework for the design process.

querying (and therefore greater ability to meet user requirements while also providing access to context-related scenario information), identifying what information the user should be given mandatorily, from a safety perspective, has not been directly measured. In Chapter 4, it was noted that protocols and checklists were featured among safety-critical domains to instill structure in agent expectations and the reception of safety information—although applying this method to the driving domain appeared to have a reduction in usability and drivers rated the interaction as largely unnecessary. This may be a unique challenge for the AV domain, as automation is currently regarded as a luxury feature rather than a safety feature. It follows that low usability, acceptance, or trust could lead to the user deactivating the feature. Therefore, pursuing knowledge on the trade-offs between legal requirements, operational safety, and user requirements are of utmost importance to allow level 3 and 4 AVs to be safely implemented onto public roads.

10.5.2   Vocal Communication in Real-World Settings

This book does not directly explore the technical challenges of applying vocal-user interfaces to automated driving. Additional challenges are introduced when passengers, complicated road scenarios, vibrations, and traffic noises are introduced. For safe operation, these interfaces require features, such as fallback inputs, in case voice is not recognized and the level of control given by vocal interaction should be considered. For example, controlling safety-critical tasks (such as overtaking) should be allocated to inputs that are accurate, and should inaccuracies occur in vocal coordination that confirmations are acquired. Use of language is also complex. Questions may not be specific, may be ambiguous, and the range of varying voice artifacts (for example dialects) makes interaction potentially difficult. This book provides promising findings for the future use of vocal and visual interfaces, although the pace at which it can be applied is in line with technological developments and addressing these design problems.

10.5.3   Validation of Concept On-Road

This book's findings are limited by their simulated nature. Focus groups, experiments, and literature searches focus on real-world implications; however, real-world settings can involve many different variables, scenarios, and road layouts that may make human–automation interaction a lot more challenging. The research approach of this book attempts to create an adaptable tool for C/HAV interaction to ameliorate these effects; however, this has not been directly addressed in the book's methodological approach. Further research will be required to factor in the complexity of real-world scenarios and test designs on-road before progressing them for manufacture.

10.5.4   A Thorough Investigation into How AV
Interaction Differs between Nations

Due to studies within this book being tailored toward UK roadways, and participants living within the UK, the findings and recommendations within this book may be tailored toward a UK road environment. As is with any major manufacturer, sales will be expected to be global. National laws, cultural values, road-user behavior, road environments, and hazards (e.g., vehicle type, weather, and road layout) will be broad and varied among various cultures. It follows that an understanding of local requirements for C/HAV interaction, particularly for the purpose of raising SA, will be of paramount importance before rolling out vehicles in certain environments. The interaction style proposed in this book does provide flexibility for various contexts; however, nations will be required to identify protocol, training, and manufacturing requirements prior to vehicle roll-out. A collaborative body of work would allow for manufacturers to implement these requirements preemptively and allow the global community to tackle these shared engineering problems collaboratively.

10.5.5   A Thorough Investigation into Demographic Variables

Within nations, population varies greatly. Demographics are an important aspect for any human-interaction design as they can drastically affect user requirements and the operability of the system. To ensure that the concepts and designs outlined within this book are influential within the public domain, trials must continue to address factors such as gender, age, socioeconomic background, and technological literacy.

10.5.6 THE ROLE OF GENDER IN VIRTUAL ASSISTANTS

A recent UN report found that virtual assistants are at risk of enhancing gender stereotypes, with many companies implementing female assistants because of their user testing (West et al., 2019). Research has shown that gender can interact with the nature of communication (i.e., what is being communicated) to influence how 'preferred' the voice is to the listener (Alesich & Rigby, 2017). However, the concept of gender within virtual assistants should not be treated with complacency, as stereotypes and bias could be further imbedded into our daily lives. Female voices as a default could reinforce aspects such as subservience (Nass et al., 1997). As virtual assistants become more commonplace, these design decisions could have great societal consequences (Nass et al., 1997). To that end, solutions should be developed and a responsibility from manufacturers must be adopted to accommodate these factors. Allowing the user to select a gender option may be a good start, even better would be the option for a nonbinary assistant. The implementation of nonbinary assistants is in the early stages of development. As the technology develops further work will be required to understand how users interact with these design concepts in the years to come.

10.5.7 C/HAV INTERACTION FOR DRIVER TRAINING AND TESTING

Section 10.1.4.8 addresses the potential for this book's findings to inform training needs for C/HAVs. The proposed 'three A's'—attention, awareness and action—only go a short distance toward understanding how driver training and testing can safely implement C/HAV operation. As AV technology progresses and becomes more widespread, understanding how to operate AV technology among other road vehicles in a wide range of situations increases. Teaching drivers to be aware of their surroundings and engage with technology correctly will have to be explored in future research, particularly when vehicles require both emergency and nonemergency handovers.

10.5.8 THE STANDARDIZATION OF AV TECHNOLOGY

As with many other safety features in a vehicle, the standardization of AV technology should include how a human interacts with the AV. As an example, physical actions for the transfer of control may have a direct impact on the safety of the vehicle operation. It follows that a thorough investigation into the most appropriate approaches to C/HAV interaction and how they can be standardized across vehicles will not only better align user mental models but also ensure that uptake and operation of C/HAVs are accessible to a wide range of users.

10.5.9 APPLYING CONCEPTS TO OTHER DOMAINS

Finally, this research intends to directly address C/HAV operation; however, due to its multidomain approach the findings within this book are not exclusive to the C/

HAV domain. Exploring how DSA, JA, and the concepts introduced in this book can be applied to preexisting and future technology will be beneficial to researchers and manufacturers.

10.5.10 CLOSING REMARKS

Individually, and collectively, chapters within this book provide a design pathway and a design solution for interaction in next-generation AVs. The future of AVs remains uncertain. It is therefore hoped that the work presented in this book influences all levels of automation interaction and can be applied to other safety-critical domains to bring more clarity to new technological developments in this domain.

A Appendix
Cue Cards for Vocal Procedure — Chapter 5

Checklist Readback:

1. Place your hands on the wheel—(read-back: My hands are on the wheel).
2. Place your foot on the accelerator—(read-back: My foot is on the accelerator).
3. There is a car lane 1/2/3—(read-back: There is a car lane 1/2/3).
4. You are in lane 1/2/3—(read-back: I am in lane 1/2/3).
5. You have 150 miles of fuel left—(read-back: I have 150 miles of fuel left).
6. You are traveling at X mph—(read-back: I am traveling at X mph).
7. Your exit is junction 14 in 5 miles—(read-back: My exit is junction 14 in 5 miles).
8. Move into the left-hand lane and exit at the next junction—(read-back: Move into the left-hand lane and exit at the next junction).
9. Please take control of the vehicle—(read-back: I have control of the vehicle).

Checklist Guided Questions:

1. Are your hands on the wheel?—Respond
2. Is your foot on the accelerator?—Respond
3. What is on your left/right?—Respond
4. Which lane are you in?—Respond
5. How much fuel do you have?—Respond '200 miles of fuel'
6. What is your speed?—Respond
7. Which exit do you need to take?—Respond 'Junction 4'
8. Which lane do you need to move into?—Respond 'left lane'
9. Please take control of the vehicle

Open Questions:

D: [Ask any questions you may have about the past, the current environment, and intentions of current driver]
A: [Respond accordingly, if unsure or not known, provide any answer you wish]

Timed:

A: **You are now required to take control in the next 60 s.** *[Begin to countdown from 60 to zero].*
D: *[Don't say anything, take control when ready]*

B Appendix
HUD Slides for Final Design Solution

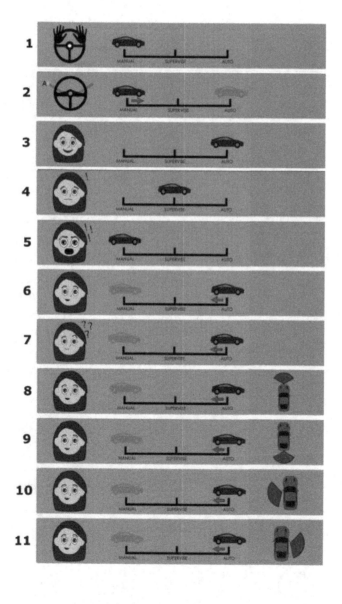

References

Aaltonen, A., Hyrskykari, A., & Räihä, K. J. (1998, January). 101 spots, or how do users read menus? In *Proceedings of the SIGCHI '98 Conference on Human Factors in Computing Systems*. New York: ACM.

Ackerman, K. A., Talleur, D. A., Carbonari, R. S., Xargay, E., Seefeldt, B. D., Kirlik, A., & Trujillo, A. C. (2017). Automation situation awareness display for a flight envelope protection system. *Journal of Guidance, Control, and Dynamics, 40*(4), 964–980.

Adamson, S., Lardner, R., & Miller, S. (1999). Safe communication at shift handover: Setting and implementing standards. In *IChemE Symposium Series 138*. Rugby: Institution of Chemical Engineers.

Alesich, S., & Rigby, M. (2017). Gendered robots: Implications for our humanoid future. *IEEE Technology and Society Magazine, 36*(2), 50–59.

Alessandrini, A., Holguín, C., & Stam, D. (2015). Automated Road Transport Systems (ARTS): The safe way to integrate automated road transport in urban areas. In *Road Vehicle Automation 2*. Cham: Springer.

Allison, C. K., & Stanton, N. A. (2020). Ideation using the "Design with Intent" toolkit: A case study applying a design toolkit to support creativity in developing vehicle interfaces for fuel-efficient driving. *Applied Ergonomics, 84*, 103026.

Anderson, J., Malone, L., Shanahan, K., & Manning, J. (2015). Nursing bedside clinical handover: An integrated review of issues and tools. *Journal of Clinical Nursing, 24*(5–6), 662–671.

Angrosino, M. V. (2016). *Naturalistic Observation*. New York: Routledge.

Annett, J. (2003). Hierarchical task analysis. In *Handbook of Cognitive Task Design*. Boca Raton, FL: CRC Press.

Arora, V. M., Johnson, J. K., Meltzer, D. O., & Humphrey, H. J. (2008). A theoretical framework and competency-based approach to improving handoffs. *Quality & Safety in Health Care, 17*(1), 11–14. doi: 10.1136/qshc.2006.018952

Arora, V. M., Reed, D. A., & Fletcher, K. E. (2014). Building continuity in handovers with shorter residency duty hours. *BMC Medical Education, 14*(1), 6. doi: 10.1186/1472-6920-14-S1-S16

Audi. (2019a). Audi Technology Portal. Retrieved from www.audi-technology-portal.de/en/electrics-electronics/driver-assistant-systems/audi-a8-audi-ai-traffic-jam-pilot. Accessed 24 January 2019.

Audi. (2019b). Audi A8 / S8 Owner's Manual. Retrieved from https://ownersmanuals2.com/audi/a8-s8-2019-owners-manual-73362. Accessed 16 June 2020.

Baddeley, A. (1992). Working memory. *Science, 255*(5044), 556–559.

Bainbridge, L. (1978). Forgotten alternatives in skill and work-load. *Ergonomics, 21*(3), 169–185.

Bainbridge, L. (1983). Ironies of automation. In *Analysis, Design and Evaluation of Man: Machine Systems*. Oxford: Pergamon.

Banks, V. A., Plant, K. L., & Stanton, N. A. (2018). Driver error or designer error: Using the perceptual cycle model to explore the circumstances surrounding the fatal Tesla crash on 7th May 2016. *Safety Science, 108*, 278–285.

Banks, V. A., & Stanton, N. A. (2016). Keep the driver in control: Automating automobiles of the future. *Applied Ergonomics, 53*, 389–395.

Barón, A., & Green, P. (2018). *Safety and Usability of Speech Interfaces for In-Vehicle Tasks while Driving: A Brief Literature Review* (Report No. UMTRI-2006–5). The University of Michigan Transportation Research Institute. Retrieved from http://umich.edu/~driving/publications/UMTRI-2006-5a.pdf

Bass, B. M., & Pak, R. (2012). Faces as ambient displays: Assessing the attention-demanding characteristics of facial expressions. In *Proceedings of the Human Factors and Ergonomics Society Annual Meeting 2012.* Thousand Oaks: Sage Publishing.

Bazilinskyy, P., & de Winter, J. (2015). Auditory interfaces in automated driving: An international survey. *PeerJ Computer Science, 1*(13).

Bazilinskyy, P., Petermeijer, S. M., Petrovych, V., Dodou, D., & De Winter, J. C. F. (2015). Take-over requests in highly automated driving: A crowdsourcing survey on auditory, vibrotactile, and visual displays. *Transportation Research Part F: Traffic Psychology and Behaviour, 56,* 82–98.

BBC. (2020). Tesla Autopilot Crash Driver "Was Playing Video Game". Retrieved from www.bbc.co.uk/news/technology-51645566?intlink_from_url=www.bbc.co.uk/news/topics/c90ymkd8lglt/driverless-cars&link_location=live-reporting-story. Accessed 20 May 2020.

Beller, J., Heesen, M., & Vollrath, M. (2013). Improving the driver: Automation interaction: An approach using automation uncertainty. *Human Factors, 55*(6), 1130–1141.

Berger, C. R., & Calabrese, R. J. (1974). Some explorations in initial interaction and beyond: Toward a developmental theory of interpersonal communication. *Human Communication Research, 1*(2), 99–112.

Bickmore, T., & Cassell, J. (2001, March). Relational agents: A model and implementation of building user trust. In *Proceedings of the SIGCHI Conference on Human Factors in Computing Systems.* New York: ACM.

Birrell, S. A., Young, M. S., Jenkins, D. P., & Stanton, N. A. (2012). Cognitive work analysis for safe and efficient driving. *Theoretical Issues in Ergonomics Science, 13*(4), 430–449.

Blake Group. (2020). Effective Communication Questionnaire. Retrieved from www.blake-group.com/sites/default/files/assessments/Communication_Assessment_Draft.pdf. Accessed 16 June 2020.

Bolstad, C. A., Endsley, M. R., & Hill, F. (2003). Tools for supporting team SA and collaboration in army operations. In *Collaborative Technology Alliances Conferences: Advanced Decision Architecture Conference.* Forest Hill: SA Technologies. Retrieved from www.researchgate.net/publication/275455620

Bonett, M. (2001). Personalization of Web Services: Opportunities and Challenges. Retrieved from www.ariadne.ac.uk/issue28/personalization/?ref=Sawos.Org. Accessed 16 June 2020.

Borman, S. (2004). The expectation maximization algorithm-a short tutorial. Retrieved from https://www.lri.fr/~sebag/COURS/EM_algorithm.pdf. Accessed 10 June 2021.

Borojeni, S. S., Chuang, L., Heuten, W., & Boll, S. (2016, October). Assisting drivers with ambient take-over requests in highly automated driving. In *Proceedings of the 8th International Conference on Automotive User Interfaces and Interactive Vehicular Application.* New York, NY: Association for Computing Machinery.

Borowitz, S. M., Waggoner-Fountain, L. A., Bass, E. J., & Sledd, R. M. (2008). Adequacy of information transferred at resident sign-out (in-hospital handover of care): A prospective survey. *Quality and Safety in Health Care, 17*(1), 6–10. Retrieved from http://qualitysafety.bmj.com/content/qhc/17/1/6.full.pdf. Accessed 16 June 2020.

Bost, N., Crilly, J., Patterson, E., & Chaboyer, W. (2012). Clinical handover of patients arriving by ambulance to a hospital emergency department: A qualitative study. *International Emergency Nursing, 20*(3), 133–141.

Boyd, M., Cumin, D., Lombard, B., Torrie, J., Civil, N., & Weller, J. (2014). Read-back improves information transfer in simulated clinical crises. *BMJ Quality & Safety*, *23*(12), 989–993.

Bradshaw, J. M., Feltovich, P., Johnson, M., Breedy, M., Bunch, L., Eskridge, T., & van Diggelen, J. (2009). From tools to teammates: Joint activity in human-agent-robot teams. In *International Conference on Human Centered Design*. Berlin, Heidelberg: Springer.

Bradshaw, J. M., Sierhuis, M., Acquisti, A., Feltovich, P., Hoffman, R., Jeffers, R., & Van Hoof, R. (2003). Adjustable autonomy and human-agent teamwork in practice: An interim report on space applications. In *Agent Autonomy*. Boston: Springer.

Brandenburg, S., & Skottke, E. M. (2014). Switching from manual to automated driving and reverse: Are drivers behaving more risky after highly automated driving? In *Proceedings from the IEEE 17th International Conference on Intelligent Transportation Systems*. Qingdao: IEEE.

Brazier, A., & Pacitti, B. (2008). Improving shift handover and maximising its value to the business. *IChemE Symposium Series*, *154*, 1–13. Retrieved from www.icheme.org/. Accessed 21 June 2020.

Brennan, P. M., & Adelman, J. (2008). U.S. Patent No. 7,333,933. Washington, DC: U.S. Patent and Trademark Office.

Brickman, B. J., Hettinger, L. J., & Haas, M. W. (2000). Multisensory interface design for complex task domains: Replacing information overload with meaning in tactical crew stations. *International Journal of Aviation Psychology*, *10*(3), 273–290.

Brooke, J. (1996). SUS-A quick and dirty usability scale. *Usability Evaluation in Industry*, *189*(194), 4–7.

Brown, J. P. (2004). Closing the communication loop: Using readback/hearback to support patient safety. *Joint Commission Journal on Quality and Safety*, *30*(8), 460–464.

Bulfone, G., Sumathy, M., Grubissa, S., & Palese, A. (2012). Effective transfer of information and responsibilities with handover: A literature review. *Assistenza Infermieristica E Ricerca*, *31*(2), 91–101.

Cadillac. (2020). Discover Cadillac. Retrieved from www.cadillac.com/world-of-cadillac/innovation/super-cruise. Accessed 21 June 2020.

Catchpole, K. R., De Leval, M. R., Mcewan, A., Pigott, N., Elliott, M. J., Mcquillan, A., Macdonald, C., & Goldman, A. J. (2007). Patient handover from surgery to intensive care: Using formula 1 pit-stop and aviation models to improve safety and quality. *Pediatric Anesthesia*, *17*(5), 470–478.

Charlton, S. G., & Starkey, N. J. (2011). Driving without awareness: The effects of practice and automaticity on attention and driving. *Transportation Research Part F: Traffic Psychology and Behaviour*, *14*(6), 456–471.

Cheah, L., Amott, D. H., Pollard, J., & Watters, D. A. (2005). Electronic medical handover: Towards safer medical care. *Medical Journal of Australia*, *183*(7), 369.

Chérif, E., & Lemoine, J. F. (2019). Anthropomorphic virtual assistants and the reactions of Internet users: An experiment on the assistant's voice. *Recherche et Applications en Marketing*, *34*(1), 28–47.

Cheung, D. S., Kelly, J. J., Beach, C., Berkeley, R. P., Bitterman, R. A., Broida, R. I., Dalsey, W. C., Farley, H. L., Fuller, D. C., Garvey, D. J., Klauer, K. M., Mccullough, L. B., Patterson, E. S., Pham, J. C., Phelan, M. P., Pines, J. M., Schenkel, S. M., Tomolo, A., Turbiak, T. W., Vozenilek, J. A., Wears, R. L., & White, M. L. (2010). Improving handoffs in the emergency department. *Annals of Emergency Medicine*, *55*(2), 171–180.

Cho, J. (2018, April). Mental models and Home Virtual Assistants (HVAs). In *Extended Abstracts of the 2018 CHI Conference on Human Factors in Computing Systems*. New York, NY: Association for Computing Machinery.

Chui, M. A., & Stone, J. A. (2012). The prescription handoff in community pharmacy: A study of its form and function. *Journal of the American Pharmacists Association*, 52(6), 161–167.

Clark, H. H. (1996). *Using Language*. Cambridge: Cambridge University Press.

Clark, H. H., & Brennan, S. E. (1991). Grounding in communication. In Resnick, L., Levine, J. M., & Teasley, S. D. (Eds.), *Perspectives on Socially Shared Cognition*. Washington, DC: APA.

Clark, J. R., Stanton, N. A., & Revell, K. M. A. (2018, July). Handover assist in highly automated vehicles: How vocal communication guides visual attention. In *International Conference on Applied Human Factors and Ergonomics*. Cham: Springer.

Clark, J. R., Stanton, N. A., & Revell, K. M. A. (2019a). Conditionally and highly automated vehicle handover: A study exploring vocal communication between two drivers. *Transportation Research Part F: Psychology and Behaviour*, 65, 699–715.

Clark, J. R., Stanton, N. A., & Revell, K. M. (2019b). Identified handover tools and techniques in high-risk domains: Using distributed situation awareness theory to inform current practices. *Safety Science*, 118, 915–924.

Clark, J. R., Stanton, N. A., & Revell, K. M. (2019c). Directability, eye-gaze, and the usage of visual displays during an automated vehicle handover task. *Transportation Research Part F: Traffic Psychology and Behaviour*, 67, 29–42. https://doi.org/10.1016/j.trf.2019.10.00

Clark, J. R., Stanton, N. A., & Revell, K. M. (2020). Automated vehicle handover interface design: Focus groups with learner, intermediate and advanced drivers. *Automotive Innovation*, 3, 1–16.

Cohen, M. D., Hilligoss, B., & Amaral, A. C. K.-B. (2012). A handoff is not a telegram: An understanding of the patient is co-constructed. *Critical Care*, 16(1), 303.

Dawson, S., King, L., & Grantham, H. (2013). Review article: Improving the hospital clinical handover between paramedics and emergency department staff in the deteriorating patient. *Emergency Medicine Australasia*, 25(5), 393–405.

de Carvalho, P. V. R., Benchekroun, T. H., & Gomes, J. O. (2012). Analysis of information exchange activities to actualize and validate situation awareness during shift change-overs in nuclear power plants. *Human Factors and Ergonomics in Manufacturing & Service Industries*, 22(2), 130–144.

Department for Transport. (2015). The Pathway to Driverless Cars: Summary Report and Action Plan. Retrieved from https://assets.publishing.service.gov.uk/government/uploads/system/uploads/attachment_data/file/401562/pathway-driverless-cars-summary.pdf. Accessed 21 June 2020.

Department of the Army. (2007). Air Traffic Services Operations FM 3-04.120. Retrieved from https://fas.org/irp/doddir/army/fm3-04-120.pdf. Accessed 21 June 2020.

de Winter, J. C. F., Eisma, Y. B., Cabrall, C. D. D., Hancock, P. A., & Stanton, N. A. (2018). Situation awareness based on eye movements in relation to the task environment. *Cognition, Technology & Work*, 21, 99–111.

de Winter, J. C. F., Happee, R., Martens, M. H., & Stanton, N. A. (2014). Effects of adaptive cruise control and highly automated driving on workload and situation awareness: A review of the empirical evidence. *Transportation Research Part F: Traffic Psychology and Behaviour*, 27, 196–217.

Douglas, H. E., Raban, M. Z., Walter, S. R., & Westbrook, J. I. (2017). Improving our understanding of multi-tasking in healthcare: Drawing together the cognitive psychology and healthcare literature. *Applied Ergonomics*, 59, 45–55.

Drach-Zahavy, A., Goldblatt, H., & Maizel, A. (2015). Between standardisation and resilience: Nurses' emergent risk management strategies during handovers. *Journal of Clinical Nursing*, 24(3–4), 592–601. doi: 10.1111/jocn.12725

Drach-Zahavy, A., & Hadid, N. (2015). Nursing handovers as resilient points of care: Linking handover strategies to treatment errors in the patient care in the following shift. *Journal of Advanced Nursing, 71*(5), 1135–1145.

Durso, F., Crutchfield, J., & Harvey, C. (2007). The cooperative shift change: An illustration using air traffic control. *Theoretical Issues in Ergonomics Science, 8*(3), 213–232.

Endsley, M. R. (1995). Toward a theory of situation awareness in dynamic systems. *Human Factors, 37*(1), 32–64.

Endsley, M. R., & Kiris, O. E. (1995). The out-of-the-loop performance problem and the level of control in automation. *Human Factors, 37*, 381–394.

Eriksson, A., Banks, V., & Stanton, N. A. (2017). Transition to manual: Comparing simulator with on-road control transitions. *Accident Analysis and Prevention, 102*, 227–234. doi: 10.1016/j.aap.2017.03.011

Eriksson, A., Marcos, I. S., Kircher, K., Västfjäll, D., & Stanton, N. A. (2015). The development of a method to assess the effects of traffic situation and time pressure on driver information preferences. In Harris, D. (Ed.), *Engineering Psychology and Cognitive Ergonomics*. Cham: Springer International Publishing.

Eriksson, A., Petermeijer, S. M., Zimmermann, M., De Winter, J. C. F., Bengler, K. J., & Stanton, N. A. (2019). Rolling out the red (and green) carpet: Supporting driver decision making in automation-to-manual transitions. *IEEE Transactions on Human-Machine Systems, 49*(1), 20–31.

Eriksson, A., & Stanton, N. A. (2017a). Takeover time in highly automated vehicles: Noncritical transitions to and from manual control. *Human Factors, 59*(4), 689–705.

Eriksson, A., & Stanton, N. A. (2017b). Driving performance after self-regulated control transitions in highly automated vehicles. *Human Factors, 59*(8), 1233–1248.

Eriksson, A., & Stanton, N. A. (2017c). The chatty co-driver: A linguistics approach applying lessons learnt from aviation incidents. *Safety Science, 99*, 94–101. https://doi.org/10.1016/j.ssci.2017.05.005

Eriksson, A., & Stanton, N. A. (2018). *Driver Reactions to Automated Vehicles: A Practical Guide for Design and Evaluation.* Boca Raton, FL: CRC Press.

Eurocontrol. (2007). Selected Safety Issues for Staffing ATC Operations. Retrieved from www.eurocontrol.int/sites/default/files/article/content/documents/nm/safety/safety-selected-safety-issues-for-staffing-atc-operations.pdf. Accessed 21 June 2020.

Eurocontrol. (2012). Guidelines for the Application of European Coordination and Transfer Procedures. Retrieved from www.eurocontrol.int/sites/default/files/content/documents/nm/airspace/airspace-atmprocedures-coordination-transfer-procedures-guidelines-1.0.pdf. Accessed 21 June 2020.

Evans., L. (2004). *Traffic Safety.* Bloomfield Hills, MI: Science Serving Society.

Fagnant, D. J., & Kockelman, K. (2015). Preparing a nation for autonomous vehicles: Opportunities, barriers and policy recommendations. *Transportation Research Part A: Policy and Practice, 77*, 167–181.

Fassert, C., & Bezzina, M. (2007). Study Report on Factors Affecting Handovers. Retrieved from www.ariatm.com/documents/R_HandoverTakeover_766289.pdf. Accessed 21 June 2020.

Federal Aviation Administration. (2010). Appendix D: Standard Operating Practice (SOP) for the Transfer of Position Responsibility. Retrieved from http://tfmlearning.faa.gov/Publications/atpubs/ATC/AppdxD.html. Accessed 21 June 2020.

Federal Aviation Administration. (2015). Form FAA 7233-4: Pre-Flight Pilot Checklist and International Flight Plan. Retrieved from www.faa.gov/documentLibrary/media/Form/FAA_7233-4_PRA_07-31-2017.pdf. Accessed 21 June 2020.

Federal Aviation Administration. (2020). Pilot Training. Retrieved from www.faa.gov/pilots/training/. Accessed 23 June 2020.

Feldhütter, A., Gold, C., Schneider, S., & Bengler, K. (2017). How the duration of automated driving influences take-over performance and gaze behavior. In *Advances in Ergonomic Design of Systems, Products and Processes*. Berlin, Heidelberg: Springer.

Flemisch, F., Heesen, M., Hesse, T., Kelsch, J., Schieben, A., & Beller, J. (2012). Towards a dynamic balance between humans and automation: Authority, ability, responsibility and control in shared and cooperative control situations. *Cognition, Technology & Work, 14*(1), 3–18.

Flink, M., Hesselink, G., Pijnenborg, L., Wollersheim, H., Vernooij-Dassen, M., Dudzik-Urbaniak, E., Orrego, C., Toccafondi, G., Schoonhoven, L., Gademan, P. J., Johnson, J. K., Ohlen, G., Hansagi, H., Olsson, M., Barach, P., & European, H. R. C. (2012). The key actor: A qualitative study of patient participation in the handover process in europe. *BMJ Quality & Safety, 21*, 89–96.

Forster, Y., Naujoks, F., & Neukum, A. (2016). Your turn or my turn?: Design of a human-machine interface for conditional automation. In *Proceedings of the 8th International Conference on Automotive User Interfaces and Interactive Vehicular Applications*. New York, NY: Association for Computing Machinery.

Fuld, R. B. (2000). The fiction of function allocation, revisited. *International Journal of Human-Computer Studies, 52*(2), 217–233.

Gold, C., Happee, R., & Bengler, K. (2017). Modeling take-over performance in level 3 conditionally automated vehicles. *Accident Analysis & Prevention, 116*, 3–13. doi: 10.1016/j.aap.2017.11.009

Gold, C., Körber, M., Lechner, D., & Bengler, K. (2016). Taking over control from highly automated vehicles in complex traffic situations: The role of traffic density. *Human Factors, 58*(4), 642–652.

Gopher, D., & Kimchi, R. (1989). Engineering psychology. *Annual Review of Psychology, 40*(1), 431–455.

Gordon, M., & Findley, R. (2011). Educational interventions to improve handover in health care: A systematic review. *Medical Education, 45*(11), 1081–1089.

Gordon, R. P. (1998). The contribution of human factors to accidents in the offshore oil industry. *Reliability Engineering & System Safety, 61*(1–2), 95–108.

GOV. (2012). Driving Lessons and Learning to Drive. Retrieved from www.gov.uk/driving-lessons-learning-to-drive. Accessed 21 June 2020.

GOV. (2017). Research and Analysis: Self-Driving Cars. Retrieved from www.gov.uk/government/publications/self-driving-cars. Accessed 21 June 2020.

Grice, H. P. (1975). Logic and conversation. In Cole, C., & Morgan, J. L. (Eds.), *Speech Acts*. New York: Academic Press.

Gross, T. K., Benjamin, L. S., Stone, E., Shook, J. E., Chun, T. H., Conners, G. P., Conway, E. E., Dudley, N. C., Fuchs, S. M., Lane, N. E., Macias, C. G., Moore, B. R., Wright, J. L., & Amer Acad, P. (2016). Handoffs: Transitions of care for children in the emergency department. *Pediatrics, 138*(5), 12.

Grusenmeyer, C. (1995). Shared functional representation in cooperative tasks: The example of shift changeover. *International Journal of Human Factors in Manufacturing, 5*(2), 163–176.

Gyselinck, V., Jamet, E., & Dubois, V. (2008). The role of working memory components in multimedia comprehension. *Applied Cognitive Psychology, 22*(3), 353–374.

Haig, K. M., Sutton, S., & Whittington, J. (2006). SBAR: A shared mental model for improving communication between clinicians. *The Joint Commission Journal on Quality and Patient Safety, 32*(3), 167–175.

Hannaford, N., Mandel, C., Crock, C., Buckley, K., Magrabi, F., Ong, M., Allen, S., & Schultz, T. (2013). Learning from incident reports in the Australian medical imaging setting: Handover and communication errors. *British Journal of Radiology, 86*(1022), 11.

Hart, S. G., & Staveland, L. E. (1988). Development of NASA-TLX (task load index): Results of empirical and theoretical research. *Advances in Psychology*, *52*, 139–183.

Harvey, C., & Stanton, N. A. (2013). *Usability Evaluation for In-Vehicle Systems*. Boca Raton, FL: CRC Press.

Health and Safety Executive. (2003). How to Organise and Run Focus Groups. Retrieved from www.hse.gov.uk/stress/assets/docs/focusgroups.pdf. Accessed 21 June 2020.

Heikoop, D. D., de Winter, J. C., van Arem, B., & Stanton, N. A. (2016). Psychological constructs in driving automation: A consensus model and critical comment on construct proliferation. *Theoretical Issues in Ergonomics Science*, *17*(3), 284–303.

Hewitt, J. P., & Shulman, D. (1979). *Self and Society: A Symbolic Interactionist Social Psychology*. Boston: Allyn and Bacon.

Hobbs, A., & Australian Transport Safety Bureau. (2008). *An Overview of Human Factors in Aviation Maintenance* (Report No. AR-2008-05). Canberra City: Australian Transport Safety Bureau. Retrieved from www.atsb.gov.au/media/27818/hf_ar-2008-055.pdf. Accessed 21 June 2020.

Hollnagel, E. (1993). *Human Reliability Analysis: Context and Control*. London: Academic Press.

Hopkin, V. D. (1989). Man-machine interface problems in designing air traffic control systems. *Proceedings of the IEEE*, *77*(11), 1634–1642.

Hornof, A. J., & Halverson, T. (2002). Cleaning up systematic error in eye-tracking data by using required fixation locations. *Behavior Research Methods, Instruments, & Computers*, *34*(4), 592–604.

Horswill, M. S., & McKenna, F. P. (2004). Drivers' hazard perception ability: Situation awareness on the road. In Banbury, S., & Trembay, S. (Eds.), *A Cognitive Approach to Situation Awareness: Theory and Application*. Aldershot: Ashgate.

Horwitz, L. I., Moin, T., & Green, M. L. (2007). Development and implementation of an oral sign-out skills curriculum. *Journal of General Internal Medicine*, *22*(10), 1470–1474.

Horwitz, L. I., Schuster, K. M., Thung, S. F., Hersh, D. C., Fisher, R. L., Shah, N., Cushing, W., Nunes, J., Silverman, D. G., & Jenq, G. Y. (2012). An institution-wide handoff task force to standardise and improve physician handoffs. *BMJ Quality & Safety*, *21*(10), 863–871.

Idris, H., Enea, G., & Lewis, T. A. (2016). Function allocation between automation and human pilot for airborne separation assurance. *IFAC-PapersOnLine*, *49*(19), 25–30.

Iedema, R., Merrick, E. T., Kerridge, R., Herkes, R., Lee, B., Anscombe, M., Rajbhandari, D., Lucey, M., & White, L. (2009). Handover: Enabling learning in communication for safety (helics): A report on achievements at two hospital sites. *Med J Aust*, *190*(11), 133–S136. Retrieved from www.mja.com.au/system/files/issues/190_11_010609/ied11188_fm.pdf. Accessed 21 June 2020.

Institute of Advanced Motorists. (2016). We Make Better Riders and Drivers. Retrieved from www.iamroadsmart.com/. Accessed 24 January 2019.

Jenkins, D. P., Stanton, N. A., Salmon, P. M., Walker, G. H., & Young, M. S. (2008). Using cognitive work analysis to explore activity allocation within military domains. *Ergonomics*, *51*(6), 798–815.

Jentsch, F., Barnett, J., Bowers, C. A., & Salas, E. (1999). Who is flying this plane anyway? What mishaps tell us about crew member role assignment and air crew situation awareness. *Human Factors*, *41*(1), 1–14.

Kerr, M. P. (2002). A qualitative study of shift handover practice and function from a socio-technical perspective. *Journal of Advanced Nursing*, *37*(2), 125–134.

Khan, A. M., Bacchus, A., & Erwin, S. (2012). Policy challenges of increasing automation in driving. *IATSS Research*, *35*(2), 79–89.

Klein, G., Feltovich, P. J., Bradshaw, J. M., & Woods, D. D. (2005). Common ground and coordination in joint activity. *Organizational Simulation*, *53*, 139–184.

Klein, G., Woods, D. D., Bradshaw, J. M., Hoffman, R. R., & Feltovich, P. J. (2004). Ten challenges for making automation a "team player" in joint human-agent activity. *IEEE Intelligent Systems, 19*(6), 91–95.

Kontogiannis, T., & Malakis, S. (2013). Strategies in controlling, coordinating and adapting performance in air traffic control: Modelling "loss of control" events. *Cognition, Technology & Work, 15*(2), 153–169.

Koo, J., Kwac, J., Ju, W., Steinert, M., Leifer, L., & Nass, C. (2015). Why did my car just do that? Explaining semi-autonomous driving actions to improve driver understanding, trust, and performance. *International Journal on Interactive Design and Manufacturing, 9*(4), 269–275.

Körber, M., Gold, C., Lechner, D., & Bengler, K. (2016). The influence of age on the take-over of vehicle control in highly automated driving. *Transportation Research Part F: Traffic Psychology and Behaviour, 39*, 19–32.

Krueger, R. A., & Casey, M. A. (2002). Designing and Conducting Focus Group Interviews. Retrieved from www.eiu.edu/ihec/Krueger-FocusGroupInterviews.pdf. Accessed 21 June 2020.

Krueger, R. A., & Casey, M. A. (2015). *Focus Groups: A Practical Guide for Applied Research*. London: Sage Publications.

Kyriakidis, M., de Winter, J. C., Stanton, N., Bellet, T., van Arem, B., Brookhuis, K., & Reed, N. (2017). A human factors perspective on automated driving. *Theoretical Issues in Ergonomics Science, 41*, 1–27.

Lardner, R. (1996). *Effective Shift Handover: A Literature Review* (Report No. OTO 93 003). Edinburgh: The Keil Centre. Retrieved from www.hse.gov.uk/research/otopdf/1996/oto96003.pdf. Accessed 21 June 2020.

Lardner, R. (2006). *Improving Communication at Shift Handover: Setting and Implementing Standards*. Edinburgh: The Keil Centre. Retrieved from www.hse.gov.uk/humanfactors/topics/standards.pdf. Accessed 21 June 2020.

Large, D. R., Clark, L., Quandt, A., Burnett, G., & Skrypchuk, L. (2017). Steering the conversation: A linguistic exploration of natural language interactions with a digital assistant during simulated driving. *Applied Ergonomics, 63*, 53–61.

Lawrence, R. H., Tomolo, A. M., Garlisi, A. P., & Aron, D. C. (2008). Conceptualizing handover strategies at change of shift in the emergency department: A grounded theory study. *BMC Health Services Research, 8*(1), 256. doi: 10.1186/1472-6963-8-256

LeBaron, C., Christianson, M. K., Garrett, L., & Ilan, R. (2016). Coordinating flexible performance during everyday work: An ethnomethodological study of handoff routines. *Organization Science, 27*(3), 514–534.

Lebie, L., Rhoades, J. A., & McGrath, J. E. (1995). Interaction process in computer-mediated and face-to-face groups. *Computer Supported Cooperative Work, 4*(2), 127–152.

Le Bris, V., Barthe, B., Marquié, J. C., Kerguelen, A., Aubert, S., & Bernadou, B. (2012). Advantages of shift changeovers with meetings: Ergonomic analysis of shift supervisors' activity in aircraft building. *Applied Ergonomics, 43*(2), 447–454.

Lee, J. D., & See, K. A. (2004). Trust in automation: Designing for appropriate reliance. *Human Factors, 46*(1), 50–80.

Lewis, P. M., & Swaim, D. J. (1988). Effect of a 12-hour/day shift on performance. In *Conference Record for 1988 IEEE Fourth Conference on Human Factors and Power Plants*. Monterey, CA: IEEE.

Li, X., Mckee, D. J., Horberry, T., & Powell, M. S. (2011). The control room operator: The forgotten element in mineral process control. *Minerals Engineering, 24*(8), 894–902.

Li, X., Powell, M. S., & Horberry, T. (2012). Human factors in control room operations in mineral processing: Elevating control from reactive to proactive. *Journal of Cognitive Engineering and Decision Making, 6*(1), 88–111.

Lockton, D. (2015). Design with Intent. Retrieved from http://designwithintent.co.uk/. Accessed 17 June 2020.

Lockton, D., Harrison, D., & Stanton, N. A. (2010). The design with intent method: A design tool for influencing user behaviour. *Applied Ergonomics, 41*(3), 382–392.

Lockton, D., & Stanton, N. A. (2010). *Design with Intent: 101 Patterns for Influencing Behaviour through Design.* Windsor: Equifine.

Louw, T., Merat, N., & Jamson, H. (2015). Engaging with highly automated driving: To be or not to be in the loop? *Iowa Research Online.* Retrieved from https://ir.uiowa.edu/cgi/viewcontent.cgi?article=1570&context=drivingassessment. Accessed 23 June 2020.

Lugano, G. (2017, May). Virtual assistants and self-driving cars. In *Proceedings of the 15th International Conference on ITS Telecommunications,* 1–5. Nis, Yugoslavia: IEEE.

Macleod, C. M., Gopie, N., Hourihan, K. L., Neary, K. R., & Ozubko, J. D. (2010). The production effect: Delineation of a phenomenon. *Journal of Experimental Psychology: Learning, Memory, and Cognition, 36*(3), 671.

Maurer, M., Gerdes, J. C., Lenz, B., & Winner, H. (2016). *Autonomous Driving.* Berlin, Germany: Springer.

Mayhew, D. R., & Simpson, H. M. (1995). *The Role of Driving Experience: Implications for Training and Licensing of New Drivers.* Canada, Toronto: Insurance Bureau of Canada.

McCall, R., McGee, F., Meschtscherjakov, A., Louveton, N., & Engel, T. (2016). Towards a taxonomy of autonomous vehicle handover situations. In *Proceedings of the 8th International Conference on Automotive User Interfaces and Interactive Vehicular Applications.* New York, NY: Association for Computing Machinery.

McHugh, M. L. (2012). Interrater reliability: The kappa statistic. *Biochemia Medica, 22*(3), 276–282.

Merat, N., & De Waard, D. (2014). Human factors implications of vehicle automation: Current understanding and future directions. *Transportation Research Part F: Traffic Psychology and Behaviour, 27,* 193–195.

Merat, N., & Jamson, A. H. (2009). Is drivers' situation awareness influenced by a fully automated driving scenario. In *Human Factors, Security and Safety.* The Netherlands: Shaker Publishing.

Merat, N., Jamson, A. H., Lai, F. C. H., & Carsten, O. (2012). Highly automated driving, secondary task performance, and driver state. *Human Factors, 54*(5), 762–771.

Merat, N., Jamson, A. H., Lai, F. C. H., Daly, M., & Carsten, O. M. J. (2014). Transition to manual: Driver behaviour when resuming control from a highly automated vehicle. *Transportation Research Part F: Traffic Psychology and Behaviour, 27,* 274–282. Retrieved from www.sciencedirect.com/science/article/pii/S1369847814001284. Accessed 21 June 2020.

Merat, N., & Lee, J. D. (2012). Preface to the special section on human factors and automation in vehicles: Designing highly automated vehicles with the driver in mind. *Human Factors, 54*(5), 681–686.

Mirnig, A. G., Gärtner, M., Laminger, A., Meschtscherjakov, A., Trösterer, S., Tscheligi, M., & McGee, F. (2017). Control transition interfaces in semiautonomous vehicles: A categorization framework and literature analysis. In *Proceedings of the 9th International Conference on Automotive User Interfaces and Interactive Vehicular Applications.* New York, NY: Association for Computing Machinery.

Mok, B., Johns, M., Lee, K. J., Miller, D., Sirkin, D., Ive, P., & Ju, W. (2015). Emergency, automation off: Unstructured transition timing for distracted drivers of automated vehicles. In *Proceedings of the IEEE 18th International Conference.* Monterey, CA: IEEE.

Molesworth, B. R., & Estival, D. (2015). Miscommunication in general aviation: The influence of external factors on communication errors. *Safety Science, 73,* 73–79.

Møller, A. M., & Myles, P. S. (2016, October). What makes a good systematic review and meta-analysis? *BJA: British Journal of Anaesthesia, 117*(4), 428–430.

Monk, A. (2003). Common ground in electronically mediated communication: Clark's theory of language use. In Carroll, J. M. (Ed.), *HCI Models, Theories, and Frameworks: Toward a Multidisciplinary Science*. Netherlands, Amsterdam: Elsevier Inc.

Morgan, P., Alford, C., & Parkhurst, G. (2016). *Handover Issues in Autonomous Driving: A Literature Review*. Project Report. Bristol: University of the West of England. Retrieved from http://eprints.uwe.ac.uk/29167. Accessed 21 June 2020.

Nass, C., Moon, Y., & Green, N. (1997). Are machines gender neutral? Gender-stereotypic responses to computers with voices. *Journal of Applied Social Psychology, 27*(10), 864–876.

Naujoks, F., Forster, Y., Wiedemann, K., & Neukum, A. (2017). A human-machine interface for cooperative highly automated driving. In *Advances in Human Aspects of Transportation*. Berlin: Springer.

Naujoks, F., Hergeth, S., Wiedemann, K., Schömig, N., Forster, Y., & Keinath, A. (2019). Test procedure for evaluating the human–machine interface of vehicles with automated driving systems. *Traffic Injury Prevention, 20*(suppl. 1), S146–S151.

Naujoks, F., Mai, C., & Neukum, A. (2014). The effect of urgency of take-over requests during highly automated driving under distraction conditions. *Advances in Human Aspects of Transportation, 7*, 431.

Naujoks, F., & Neukum, A. (2014). Specificity and timing of advisory warnings based on cooperative perception. In *Proceedings of Mensch & Computer 2014*. Berlin: De Gruyter Oldenbourg.

Neisser, U. (1976). *Cognition and Reality: Principles and Implications of Cognitive Psychology*. New York, NY: W. H. Freeman/Times Books/Henry Holt & Co.

NHTSA. (2013). U.S. Department of Transportation Releases Policy on Automated Vehicle Development. Retrieved from www.transportation.gov/briefing-room/us-department-transportation-releases-policy-automated-vehicle-development. Accessed 21 June 2020.

Nolan, M. (2010). *Fundamentals of Air Traffic Control* (5th ed.). Clifton Park: Thomson Delmar Learning.

Norman, D. A. (2015). The human side of automation. In *Road Vehicle Automation 2*. Cham: Springer.

Norris, B., West, J., Anderson, O., Davey, G., & Brodie, A. (2014). Taking ergonomics to the bedside: A multi-disciplinary approach to designing safer healthcare. *Applied Ergonomics, 45*(3), 629–638.

NSAI. (2018). *Ergonomics of Human-System Interaction, Part 11: Usability: Definitions and Concepts* (ISO 9241–11:2018). Retrieved from https://infostore.saiglobal.com/preview/is/en/2018/i.s.eniso9241-11-2018.pdf?sku=1980667. Accessed 23 June 2020.

Nwiabu, N., & Adeyanju, I. (2012). User centred design approach to situation awareness. *International Journal of Computer Applications, 49*(17), 26–30.

Östlund, U., Kidd, L., Wengström, Y., & Rowa-Dewar, N. (2011). Combining qualitative and quantitative research within mixed method research designs: A methodological review. *International Journal of Nursing Studies, 48*(3), 369–383.

Oviatt, S. (1997). Multimodal interactive maps: Designing for human performance. *Human-Computer Interaction, 12*(1), 93–129.

Owsley, C., & McGwin, G., Jr. (2010). Vision and driving. *Vision Research, 50*(23), 2348–2361.

The Oxford English Dictionary. (2018a). Customize. Retrieved from https://en.oxford dictionaries.com/definition/customize. Accessed 24 January 2019.

The Oxford English Dictionary. (2018b). Personalize. Retrieved from https://en.oxford dictionaries.com/definition/personalize. Accessed 24 January 2019.

Parke, B., & Mishkin, A. (2005). Best practices in shift handover communication: Mars exploration rover surface operations. In *Proceedings of the International Association for the Advancement of Space Safety Conference*. Noordwijk: IAASS.

Parke, B., Hobbs, A., & Kanki, B. (2010). Passing the baton: An experimental study of shift handover. In *Proceedings of the Human Factors and Ergonomics Society Annual Meeting*. Los Angeles: SAGE Publications.

Parke, B., & Kanki, B. G. (2008). Best practices in shift turnovers: Implications for reducing aviation maintenance turnover errors as revealed in asrs reports. *The International Journal of Aviation Psychology*, 18(1), 72–85. doi: 10.1080/10508410701749464

Paté-Cornell, M. E. (1993). Learning from the piper alpha accident: A postmortem analysis of technical and organizational-factors. *Risk Analysis*, 13(2), 215–232.

Patterson, E. S. (2008). Structuring flexibility: The potential good, bad and ugly in standardisation of handovers. *Qual Saf Health Care*, 17(1), 4–5.

Patterson, E. S., Roth, E. M., Woods, D. D., Chow, R., & Gomes, J. O. (2004). Handoff strategies in settings with high consequences for failure: Lessons for health care operations. *International Journal for Quality in Health Care*, 16(2), 125–132.

Patterson, E. S., & Woods, D. D. (2001). Shift changes, updates, and the on-call architecture in space shuttle mission control. *Computer Supported Cooperative Work*, 10(3–4), 317–346.

Paxion, J., Galy, E., & Berthelon, C. (2014). Mental workload and driving. *Frontiers in Psychology*, 5, 1344.

Petermeijer, S. M., Bazilinskyy, P., Bengler, K., & de Winter, J. (2017a). Take-over again: Investigating multimodal and directional TORs to get the driver back into the loop. *Applied Ergonomics*, 62, 204–215.

Petermeijer, S. M., Hornberger, P., Ganotis, I., de Winter, J. C., & Bengler, K. J. (2017b). The design of a vibrotactile seat for conveying take-over requests in automated driving. In *Advances in Human Aspects of Transportation: AHFE 2017*. Cham: Springer.

Philibert, I. (2009). Use of strategies from high-reliability organisations to the patient handoff by resident physicians: Practical implications. *Quality & Safety in Health Care*, 18(4), 261–266.

Plant, K. L., & Stanton, N. A. (2012). Why did the pilots shut down the wrong engine? Explaining errors in context using Schema Theory and the perceptual cycle model. *Safety Science*, 50(2), 300–315.

Plant, K. L., & Stanton, N. A. (2013). What is on your mind? Using the perceptual cycle model and critical decision method to understand the decision making process in the cockpit. *Ergonomics*, 56(8), 1232–1250.

Plocher, T., Yin, S., Laberge, J., Thompson, B., & Telner, J. (2011). Effective shift handover. In Harris, D. (Ed.), *Lecture Notes in Computer Science*. Heidelberg: Springer.

Politis, I., Brewster, S. A., & Pollick, F. (2014). Evaluating multimodal driver displays under varying situational urgency. In *Proceedings of the SIGCHI Conference on Human Factors in Computing Systems*. Canada, Toronto: ACM.

Politis, I., Brewster, S. A., & Pollick, F. (2015). Language-based multimodal displays for the handover of control in autonomous cars. In *Proceedings of the 7th International Conference on Automotive User Interfaces and Interactive Vehicular Applications*. New York, NY: Association for Computing Machinery.

Ponsa, P., Vilanova, R., & Amante, B. (2009). Towards integral human-machine system conception: From automation design to usability concerns. In *Human System Interactions 2009*. Monterey, CA: IEEE.

Pucher, P. H., Johnston, M. J., Aggarwal, R., Arora, S., & Darzi, A. (2015). Effectiveness of interventions to improve patient handover in surgery: A systematic review. *Surgery*, 158(1), 85–95.

Quimby, A. R., Maycock, G., Carter, I. D., Dixon, R., & Wall, J. (1986). *Perceptual Abilities of Accident Involved Drivers*. Wokingham: Transport and Road Research Laboratory.

Raduma-Tomas, M. A., Flin, R., Yule, S., & Williams, D. (2011). Doctors' handovers in hospitals: A literature review. *BMJ Quality & Safety, 20*(2), 128–133.

Randell, R., Wilson, S., Woodward, P., & Galliers, J. (2011). The ConStratO model of handover: A tool to support technology design and evaluation. *Behaviour & Information Technology, 30*(4), 489–498.

Raptis, D. A., Fernandes, C., Chua, W., & Boulos, P. B. (2009). Electronic software significantly improves quality of handover in a London teaching hospital. *Health Informatics Journal, 15*(3), 191–198.

Rasmussen, J., Pejtersen, A. M., & Schmidt, K. (1990). *Taxonomy for Cognitive Work Analysis*. Roskilde: Riso National Laboratory.

Rayo, M. F., Mount-Campbell, A. F., O'brien, J. M., White, S. E., Butz, A., Evans, K., & Patterson, E. S. (2014). Interactive questioning in critical care during handovers: A transcript analysis of communication behaviours by physicians, nurses and nurse practitioners. *BMJ Quality & Safety, 23*(6), 483–489.

Read, G. J., Salmon, P. M., Lenné, M. G., & Jenkins, D. P. (2015). Designing a ticket to ride with the cognitive work analysis design toolkit. *Ergonomics, 58*(8), 1266–1286.

Reeves, B., & Nass, C. (1996). *How People Treat Computers, Television, and New Media Like Real People and Places*. Stanford: CSLI Publications.

Revell, K. M. A., Richardson, J., Langdon, P., Bradley, M., Politis, I., Thompson, S., Skrypchuck, L., O' Donoghue, J., Moszakitis, A., & Stanton, N. A. (2020). Breaking the cycle of frustration: Applying Neisser's perceptual cycle model to drivers of semi-autonomous vehicles. *Applied Ergonomics, 85*, 103037.

Revell, K. M. A., & Stanton, N. A. (2012). Models of models: Filtering and bias rings in depiction of knowledge structures and their implications for design. *Ergonomics, 55*(9), 1073–1092.

Riesenberg, L. A. (2012). Shift-to-shift handoff research: Where do we go from here? *Journal of Graduate Medical Education, 4*(1), 4–8.

Riesenberg, L. A., Leitzsch, J., & Little, B. W. (2009a). Systematic review of handoff mnemonics literature. *American Journal of Medical Quality, 24*(3), 196–204.

Riesenberg, L. A., Leitzsch, J., Massucci, J. L., Jaeger, J., Rosenfeld, J. C., Patow, C., Padmore, J. S., & Karpovich, K. P. (2009b). Residents' and attending physicians' handoffs: A systematic review of the literature. *Academic Medicine, 84*(12), 1775–1787.

The Royal College of Surgeons of England. (2007). *Safe Handover: Guidance from the Working Time Directive Working Party*. London: RSENG. Retrieved from www.rcseng.ac.uk/-/media/files/rcs/library-and-publications/non-journal-publications/safe-handovers.pdf. Accessed 21 June 2020.

The Royal Society for the Prevention of Accidents. (2018). Accidents Don't Have to Happen. Retrieved from www.rospa.com/. Accessed 24 January 2019.

SAE J3016 On-Road Automated Vehicles Standards Committee. (2016). Taxonomy and Definitions for Terms Related to On-Road Motor Vehicle Automated Driving Systems. Retrieved from http://standards.sae.org/j3016_201401/. Accessed 21 June 2020.

Salas, E., Burke, C. S., & Cannon-Bowers, J. A. (2000). Teamwork: Emerging principles. *International Journal of Management Reviews, 2*(4), 339–356.

Salas, E., Dickinson, T. L., Converse, S. A., & Tannenbaum, S. I. (1992). *Toward an Understanding of Team Performance and Training*. New York, NY: Ablex Publishing.

Salmon, P. M., Regan, M., Lenné, M. G., Stanton, N. A., & Young, K. (2007). Work domain analysis and intelligent transport systems: Implications for vehicle design. *International Journal of Vehicle Design, 45*(3), 426–448.

Salmon, P. M., Stanton, N. A., & Jenkins, D. P. (2009). *Distributed Situation Awareness: Theory, Measurement and Application to Teamwork*. Boba Raton, FL: CRC Press.

Salmon, P. M., Walker, G. H., & Stanton, N. A. (2016). Pilot error versus sociotechnical systems failure: A distributed situation awareness analysis of Air France 447. *Theoretical Issues in Ergonomics Science, 17*(1), 64–79.

Sanders, E. B. N. (2003). From user-centered to participatory design approaches. In *Design and the Social Sciences*. Boba Raton, FL: CRC Press.

Sarter, N. B., & Woods, D. D. (1992). Mode error in supervisory control of automated systems. In *Proceedings of the Human Factors and Ergonomics Society Annual Meeting 1992*. Los Angeles, CA: SAGE Publications.

Sarter, N. B., & Woods, D. D. (1995). How in the world did we ever get into that mode? Mode error and awareness in supervisory control. *Human Factors, 37*(1), 5–19.

Sarter, N. B., Woods, D. D., & Billings, C. E. (1997). Automation surprises. *Handbook of Human Factors and Ergonomics, 2*, 1926–1943.

Schieben, A., Temme, G., Köster, F., & Flemisch, F. (2011). How to interact with a highly automated vehicle: Generic interaction design schemes and test results of a usability assessment. *Human Centred Automation*, 251–266.

Seppelt, B. D., & Lee, J. D. (2019). Keeping the driver in the loop: Dynamic feedback to support appropriate use of imperfect vehicle control automation. *International Journal of Human-Computer Studies, 125*, 66–80.

Sheridan, T. B. (2002). *Humans and Automation: System Design and Research Issues*. Hoboken, NJ: John Wiley & Sons.

Singer, J. I., & Dean, J. (2006). Emergency physician intershift handovers: An analysis of our transitional care. *Pediatric Emergency Care, 22*(10), 751–754.

Small, E., Finn, B., & Adams, D. (2011). *U.S. Patent No. 8,078,359*. Washington, DC: Patent and Trademark Office.

Sorensen, L. J., & Stanton, N. A. (2016). Keeping it together: The role of transactional situation awareness in team performance. *International Journal of Industrial Ergonomics, 53*, 267–273.

Spinuzzi, C. (2011). Losing by expanding: Corralling the runaway object. *Journal of Business and Technical Communication, 25*(4), 449–486.

Spooner, A. J., Corley, A., Chaboyer, W., Hammond, N. E., & Fraser, J. F. (2015). Measurement of the frequency and source of interruptions occurring during bedside nursing handover in the intensive care unit: An observational study. *Australian Critical Care, 28*(1), 19–23.

Staggers, N., & Blaz, J. W. (2013). Research on nursing handoffs for medical and surgical settings: An integrative review. *Journal of Advanced Nursing, 69*(2), 247–262.

Stanton, N. A. (1993). *Human Factors in Nuclear Safety*. London: Taylor & Francis.

Stanton, N. A. (2015). More responses to autonomous vehicles. *Ingenia, 62*, 9–10.

Stanton, N. A., & Allison, C. K. (2020). Driving towards a greener future: An application of cognitive work analysis to promote fuel-efficient driving. *Cognition, Technology & Work, 22*(1), 125–142.

Stanton, N. A., & Ashleigh, M. (2000). A field study of team working in a new human supervisory control system. *Ergonomics, 43*(8), 1190–1209.

Stanton, N. A., Dunoyer, A., & Leatherland, A. (2011). Detection of new in-path targets by drivers using stop & go adaptive cruise control. *Applied Ergonomics, 42*(4), 592–601.

Stanton, N. A., & Edworthy, J. (1999). *Human Factors in Auditory Warnings*. Farnham: Ashgate.

Stanton, N. A., & Marsden, P. (1996). From fly-by-wire to drive-by-wire: Safety implications of automation in vehicles. *Safety Science, 24*(1), 35–49.

Stanton, N. A., Salmon, P. M., Jenkins, D. P., & Walker, G. H. (2010). *Human Factors in the Design and Evaluation of Central Control Room Operations*. Boca Raton: CRC Press.

Stanton, N. A., Salmon, P. M., Rafferty, L. A., Walker, G. H., Baber, C., & Jenkins, D. P. (2017a). *Human Factors Methods: A Practical Guide for Engineering and Design*. Boba Raton: CRC Press.

Stanton, N. A., Salmon, P. M., Walker, G. H., Hancock, P. A., & Salas, E. (2017b). State-of-science: Situation awareness in individuals, teams and systems. *Ergonomics*, *60*(4), 449–466.

Stanton, N. A., Salmon, P. M., Walker, G. H., & Stanton, M. (2019). Models and methods for collision analysis: A comparison study based on the uber collision with a pedestrian. *Safety Science*, *120*, 117–128.

Stanton, N. A., Stewart, R., Harris, D., Houghton, R. J., Baber, C., McMaster, R., & Linsell, M. (2006). Distributed situation awareness in dynamic systems: Theoretical development and application of an ergonomics methodology. *Ergonomics*, *49*, 12–13, 1288–1311.

Stanton, N. A., Walker, G. H., Young, M. S., Kazi, T. A., & Salmon, P. M. (2007). Changing drivers' minds: The evaluation of an advanced driver coaching system. *Ergonomics*, *50*(8), 1209–1234.

Stanton, N. A., & Young, M. S. (2000). A proposed psychological model of driving automation. *Theoretical Issues in Ergonomics Science*, *1*(4), 315–331.

Stanton, N. A., Young, M. S., & McCaulder, B. (1997). Drive-by-wire: The case of driver workload and reclaiming control with adaptive cruise control. *Safety Science*, *27*(2), 149–159.

Stanton, N. A., Young, M. S., Walker, G. H., Turner, H., & Randle, S. (2001). Automating the driver's control tasks. *International Journal of Cognitive Ergonomics*, *5*(3), 221–236.

Starmer, A. J., O'toole, J. K., Rosenbluth, G., Calaman, S., Balmer, D., West, D. C., Bale, J. F., Jr., Yu, C. E., Noble, E. L., Tse, L. L., Srivastava, R., Landrigan, C. P., Sectish, T. C., & Spector, N. D. (2014b). Development, implementation, and dissemination of the i-pass handoff curriculum: A multisite educational intervention to improve patient handoffs. *Academic Medicine*, *89*(6), 876–884.

Starmer, A. J., Spector, N. D., Srivastava, R., West, D. C., Rosenbluth, G., Allen, A. D., Noble, E. L., Tse, L. L., Dalal, A. K., Keohane, C. A., Lipsitz, S. R., Rothschild, J. M., Wien, M. F., Yoon, C. S., Zigmont, K. R., Wilson, K. M., O'toole, J. K., Solan, L. G., Aylor, M., Bismilla, Z., Coffey, M., Mahant, S., Blankenburg, R. L., Destino, L. A., Everhart, J. L., Patel, S. J., Bale, J. F. J., Spackman, J. B., Stevenson, A. T., Calaman, S., Cole, F. S., Balmer, D. F., Hepps, J. H., Lopreiato, J. O., Yu, C. E., Sectish, T. C., & Landrigan, C. P. (2014a). Changes in medical errors after implementation of a handoff program. *New England Journal of Medicine*, *371*(19), 1.

Street, M., Eustace, P., Livingston, P. M., Craike, M. J., Kent, B., & Patterson, D. (2011). Communication at the bedside to enhance patient care: A survey of nurses' experience and perspective of handover. *International Journal of Nursing Practice*, *17*(2), 133–140.

Sun, Q. C., Xia, J. C., Falkmer, T., & Lee, H. (2016). Investigating the spatial pattern of older drivers' eye fixation behaviour and associations with their visual capacity. *Journal of Eye Movement Research*, *9*(6), 1–16.

Sutcliffe, K. M., Lewton, E., & Rosenthal, M. M. (2004). Communication failures: An insidious contributor to medical mishaps. *Academic Medicine*, *79*(2), 186–194.

Tesla. (2018). Tesla Model S. Retrieved from www.tesla.com/models. Accessed 21 June 2020.

Tesla. (2020). Model S. Owner's Manual. Retrieved from www.tesla.com/sites/default/files/model_s_owners_manual_north_america_en_us.pdf

Thomas, M. J., Schultz, T. J., Hannaford, N., & Runciman, W. B. (2013). Failures in transition: Learning from incidents relating to clinical handover in acute care. *Journal for Healthcare Quality*, *35*(3), 49–56.

Tobiano, G., Chaboyer, W., & Mcmurray, A. (2013). Family members' perceptions of the nursing bedside handover. *Journal of Clinical Nursing, 22*(1–2), 192–200.

Tonnis, M., Sandor, C., Klinker, G., Lange, C., & Bubb, H. (2005, October). Experimental evaluation of an augmented reality visualization for directing a car driver's attention. In *Fourth IEEE and ACM International Symposium on Mixed and Augmented Reality,* 56–59. Vienna: IEEE.

Underwood, G., Crundall, D., & Chapman, P. (2011). Driving simulator validation with hazard perception. *Transportation Research Part F: Traffic Psychology and Behaviour, 14*(6), 435–446.

U.S. Department of Transportation. (1995). Positive Exchange of Flight Controls Program. Retrieved from https://ntl.bts.gov/lib/1000/1100/1125/ac61-115.pdf. Accessed 21 June 2020.

van der Laan, J. D., Heino, A., & De Waard, D. (1997). A simple procedure for the assessment of acceptance of advanced transport telematics. *Transportation Research, Part C: Emerging Technologies, 5*(1), 1–10.

van Rensen, E. L., Groen, E. S. T., Numan, S. C., Smit, M. J., Cremer, O. L., Tates, K., & Kalkman, C. J. (2012). Multitasking during patient handover in the recovery room. *Anesthesia & Analgesia, 115*(5), 1183–1187.

van Sluisveld, N., Zegers, M., Westert, G., Van Der Hoeven, J. G., & Wollersheim, H. (2013). A strategy to enhance the safety and efficiency of handovers of ICU patients: Study protocol of the picup study. *Implementation Science, 8*(67), 1–9.

van Wijk, R., Jansen, J. J., & Lyles, M. A. (2008). Inter-and intra-organizational knowledge transfer: A meta-analytic review and assessment of its antecedents and consequences. *Journal of Management Studies, 45*(4), 830–853.

Venkatesh, V., & Bala, H. (2008). Technology acceptance model 3 and a research agenda on interventions. *Decision Sciences, 39*(2), 273–315.

Verberne, F. M., Ham, J., & Midden, C. J. (2012). Trust in smart systems: Sharing driving goals and giving information to increase trustworthiness and acceptability of smart systems in cars. *Human Factors, 54*(5), 799–810.

Vicente, K. J. (1999). *Cognitive Work Analysis: Toward Safe, Productive, and Healthy Computer-Based Work.* Boca Roton: CRC Press.

Vitense, H. S., Jacko, J. A., & Emery, V. K. (2003). Multimodal feedback: An assessment of performance and mental workload. *Ergonomics, 46*(1–3), 68–87.

Volvo Car Group. (2015). Volvo Cars Reveals Safe and Seamless User Interface for Self-Driving-Cars. Retrieved from www.media.volvocars.com/global/en-gb/media/press releases/167739/volvo-cars-reveals-safe-and-seamless-user-interface-for-self-driving-cars. Accessed 21 June 2020.

Wachter, R., & Shojania, K. G. (2004). *Internal Bleeding: The Truth Behind America's Terrifying Epidemic of Medical Mistakes.* New York, NY: Rugged Land.

Walch, M., Lange, K., Baumann, M., & Weber, M. (2015). Autonomous driving: Investigating the feasibility of car-driver handover assistance. In *Proceedings of the 7th International Conference on Automotive User Interfaces and Interactive Vehicular Applications.* New York, NY: Association for Computing Machinery.

Walch, M., Mühl, K., Kraus, J., Stoll, T., Baumann, M., & Weber, M. (2017). From car-driver-handovers to cooperative interfaces: Visions for driver: Vehicle interaction in automated driving. In *Automotive User Interfaces.* Berlin: Springer.

Waldrop, M. M. (2015). No drivers required. *Nature, 518*(7537), 20–24.

Walker, G. H., Stanton, N. A., Baber, C., Wells, L., Gibson, H., Salmon, P., & Jenkins, D. (2010). From ethnography to the EAST method: A tractable approach for representing distributed cognition in air traffic control. *Ergonomics, 53*(2), 184–197.

Walker, G. H., Stanton, N. A., Kazi, T. A., Salmon, P. M., & Jenkins, D. P. (2009). Does advanced driver training improve situational awareness? *Applied Ergonomics*, *40*(4), 678–687.

Walker, G. H., Stanton, N. A., & Salmon, P. (2016). Trust in vehicle technology. *International Journal of Vehicle Design*, *70*(2), 157–182.

Ward, N. J. (2000). Automation of task processes: An example of intelligent transportation systems. *Human Factors and Ergonomics in Manufacturing & Service Industries*, *10*(4), 395–408.

Waytz, A., Heafner, J., & Epley, N. (2014). The mind in the machine: Anthropomorphism increases trust in an autonomous vehicle. *Journal of Experimental Social Psychology*, *52*, 113–117.

Weikert, C., & Johansson, C. R. (1999). Analysing incident reports for factors contributing to air traffic control related incidents. In *Proceedings of the Human Factors and Ergonomics Society 43rd Annual Meeting*. Santa Monica: Human Factors and Ergonomics Soc.

Weld, D., Anderson, C., Domingos, P., Etzioni, O., Gajos, K. Z., Lau, T., & Wolfman, S. (2003). Automatically personalizing user interfaces. In *Proceedings of the 18th International Joint Conference on Artificial Intelligence*. San Francisco, CA: ACM.

West, M., Kraut, R., & Chew, H. E. (2019). I'd blush if I could: Closing gender divides in digital skills through education. *EQUALS and UNESCO*. Retrieved from https://unesdoc.unesco.org/ark:/48223/pf0000367416. Accessed 7 September 2020.

Wickens, C. D. (1991). *Multiple-Task Performance*. Washington, DC: APA.

Wilkinson, J., & Lardner, R. (2012). Pass it on! Revisiting shift handover after Buncefield. *Loss Prevention Bulletin*, *229*, 25–32.

Wilkinson, J., & Lardner, R. (2013). Shift handover after Buncefield. *Proceedings of the 14th Symposium on Loss Prevention and Safety Promotion in the Process Industries*, *31*, 295–300.

Willemsen, D., Stuiver, A., & Hogema, J. (2014). Transition of control: Automation giving back control to the driver. In *Proceedings of the 5th International Conference on Applied Human Factors and Ergonomics*. Krakow: Poland.

Yamagishi, T., & Yamagishi, M. (1994). Trust and commitment in the United States and Japan. *Motivation and Emotion*, *18*, 129–166.

Yan, X., Zhang, X., Zhang, Y., Li, X., & Yang, Z. (2016). Changes in drivers' visual performance during the collision avoidance process as a function of different field of views at intersections. *PLoS One*, *11*(10), e0164101.

Young, M. S., & Stanton, N. A. (2002a). Attention and automation: New perspectives on mental underload and performance. *Theoretical Issues in Ergonomics Science*, *3*(2), 178–194.

Young, M. S., & Stanton, N. A. (2002b). Malleable attentional resources theory: A new explanation for the effects of mental underload on performance. *Human Factors*, *44*(3), 365–375.

Young, M. S., & Stanton, N. A. (2004). Taking the load off: Investigations of how adaptive cruise control affects mental workload. *Ergonomics*, *47*(9), 1014–1035.

Young, M. S., & Stanton, N. A. (2007a). What's skill got to do with it? Vehicle automation and driver mental workload. *Ergonomics*, *50*(8), 1324–1339.

Young, M. S., & Stanton, N. A. (2007b). Miles away: Determining the extent of secondary task interference on simulated driving. *Theoretical Issues in Ergonomics Science*, *8*(3), 233–253.

Zeeb, K., Buchner, A., & Schrauf, M. (2015). What determines the take-over time? An integrated model approach of driver take-over after automated driving. *Accident Analysis & Prevention*, *78*, 212.

Index

Printed in the United States
by Baker & Taylor Publisher Services